Recent Advances in Metal, Ceramic, and Metal-Ceramic Composite Films/Coatings

Recent Advances in Metal, Ceramic, and Metal-Ceramic Composite Films/Coatings

Editor

Małgorzata Norek

MDPI • Basel • Beijing • Wuhan • Barcelona • Belgrade • Manchester • Tokyo • Cluj • Tianjin

Editor
Małgorzata Norek
Military University of
Technology
Poland

Editorial Office
MDPI
St. Alban-Anlage 66
4052 Basel, Switzerland

This is a reprint of articles from the Special Issue published online in the open access journal *Coatings* (ISSN 2079-6412) (available at: https://www.mdpi.com/journal/coatings/special_issues/Met_Ceram_Compos_Film_Coat).

For citation purposes, cite each article independently as indicated on the article page online and as indicated below:

LastName, A.A.; LastName, B.B.; LastName, C.C. Article Title. *Journal Name* **Year**, *Volume Number*, Page Range.

ISBN 978-3-0365-4597-4 (Hbk)
ISBN 978-3-0365-4598-1 (PDF)

© 2022 by the authors. Articles in this book are Open Access and distributed under the Creative Commons Attribution (CC BY) license, which allows users to download, copy and build upon published articles, as long as the author and publisher are properly credited, which ensures maximum dissemination and a wider impact of our publications.

The book as a whole is distributed by MDPI under the terms and conditions of the Creative Commons license CC BY-NC-ND.

Contents

Małgorzata Norek
Recent Advances in Metal, Ceramic, and Metal–Ceramic Composite Films/Coatings
Reprinted from: *Coatings* 2022, *12*, 572, doi:10.3390/coatings12050571 1

Sheng Hong, Yuping Wu, Bo Wang and Jinran Lin
Improvement in Tribological Properties of Cr12MoV Cold Work Die Steel by HVOF Sprayed WC-CoCr Cermet Coatings
Reprinted from: *Coatings* 2019, *9*, 825, doi:10.3390/coatings12050571 3

Mateusz Kopec, Dominik Kukla, Xin Yuan, Wojciech Rejmer, Zbigniew L. Kowalewski and Cezary Senderowski
Aluminide Thermal Barrier Coating for High Temperature Performance of MAR 247 Nickel Based Superalloy
Reprinted from: *Coatings* 2021, *11*, 48, doi:10.3390/coatings11010048 17

Xinqiang Lu, Shouren Wang, Tianying Xiong, Daosheng Wen, Gaoqi Wang and Hao Du
Study on Corrosion Resistance and Wear Resistance of Zn–Al–Mg/ZnO Composite Coating Prepared by Cold Spraying
Reprinted from: *Coatings* 2019, *9*, 505, doi:10.3390/coatings9080505 29

Jacek Tomków, Artur Czupryński and Dariusz Fydrych
The Abrasive Wear Resistance of Coatings Manufactured on High-Strength Low-Alloy (HSLA) Offshore Steel in Wet Welding Conditions
Reprinted from: *Coatings* 2020, *10*, 219, doi:10.3390/coatings10030219 45

Bingjie Mai, Youlu Chen, Ying Zhang, Yongsheng Huang, Juanli Wang, Yuhu Li, Ming Cao and Jing Cao
Analysis of Iron Anchor Diseases Unearthed from Gudu Ruins in Xianyang City, Shaanxi Province, China
Reprinted from: *Coatings* 2022, *12*, 381, doi:10.3390/coatings12030381 59

Navid Attarzadeh and Chintalapalle V. Ramana
Plasma Electrolytic Oxidation Ceramic Coatings on Zirconium (Zr) and ZrAlloys: Part I—Growth Mechanisms, Microstructure, and Chemical Composition
Reprinted from: *Coatings* 2021, *11*, 634, doi:10.3390/coatings11060634 71

Navid Attarzadeh and C. V. Ramana
Plasma Electrolytic Oxidation Ceramic Coatings on Zirconium (Zr) and Zr-Alloys: Part-II: Properties and Applications
Reprinted from: *Coatings* 2021, *11*, 620, doi:10.3390/coatings11060620 95

Stevan Stojadinović and Aleksandar Ćirić
Ce^{3+}/Eu^{2+} Doped Al_2O_3 Coatings Formed by Plasma Electrolytic Oxidation of Aluminum: Photoluminescence Enhancement by $Ce^{3+} \rightarrow Eu^{2+}$ Energy Transfer
Reprinted from: *Coatings* 2019, *9*, 819, doi:10.3390/coatings9120819 113

Maksymilian Włodarski, Matti Putkonen and Małgorzata Norek
Infrared Absorption Study of Zn–S Hybrid and ZnS Ultrathin Films Deposited on Porous AAO Ceramic Support
Reprinted from: *Coatings* 2020, *10*, 459, doi:10.3390/coatings10050459 123

Alexander Poznyak, Andrei Pligovka, Ulyana Turavets and Małgorzata Norek
On-Aluminum and Barrier Anodic Oxide: Meeting the Challenges of Chemical Dissolution Rate in Various Acids and Solutions
Reprinted from: *Coatings* **2020**, *10*, 875, doi:10.3390/coatings10090875 . . . . **135**

Editorial

Recent Advances in Metal, Ceramic, and Metal–Ceramic Composite Films/Coatings

Małgorzata Norek

Institute of Materials Science and Engineering, Faculty of Advanced Technologies and Chemistry, Military University of Technology, Str. Gen. S. Kaliskiego 2, 00-908 Warsaw, Poland; malgorzata.norek@wat.edu.pl

Citation: Norek, M. Recent Advances in Metal, Ceramic, and Metal–Ceramic Composite Films/Coatings. *Coatings* 2022, 12, 572. https://doi.org/10.3390/coatings12050571

Received: 19 April 2022
Accepted: 20 April 2022
Published: 22 April 2022

Publisher's Note: MDPI stays neutral with regard to jurisdictional claims in published maps and institutional affiliations.

Copyright: © 2022 by the author. Licensee MDPI, Basel, Switzerland. This article is an open access article distributed under the terms and conditions of the Creative Commons Attribution (CC BY) license (https://creativecommons.org/licenses/by/4.0/).

Coating materials and technologies are becoming increasingly important in many research areas because they can provide an efficient and affordable way to engineer materials with desirable properties for a broad range of applications. Depending on the method used, coatings can induce significant structural and microstructural changes in a given material. However, systematic research is necessary to fully understand these changes for a more rational design of a material with desirable properties. The range of properties that can be tailored include mechanical (e.g., hardness, fatigue limit, elongation, tensile strength), thermal (e.g., thermal conductivity, thermal expansion), electrical (e.g., electrical conductivity, dielectric strength), and many other functional properties, including corrosion resistance, hydrophobicity, wettability, optical properties, etc. This Special Issue gathers those studies that show the great potential of a diverse range of metal, ceramic, and metal–ceramic composite coatings for use in materials engineering.

The wear resistance of cold work die steel was significantly improved after the application of high-velocity, oxygen-fuel (HVOF)-sprayed WC-CoCr coatings, irrespective of the test temperature. This better sliding wear behavior was ascribed to the presence of nanocrystalline grains and the fcc-Co phase in the coating [1]. The high-temperature fatigue (at 900 °C) and creep performance of a MAR 247 nickel-based superalloy was also improved after an aluminide coating with the use of the chemical vapor deposition (CVD) process [2]. Two types of Zn-Al-Mg/ZnO composite coatings were deposited on a Q235 metallic substrate by the cold-spray technique in the study of [3]. After the coating was applied, the base material demonstrated better friction, wear, and corrosion resistance. Additionally, photocatalytic properties of the coated substrate were studied and showed a better photocatalytic efficiency in terms of the methyl blue degradation compared to that of the uncoated Q235.

The investigation of various conditions of coating preparation and performance is very important with respect to their subsequent exploitation or reparation in aggressive environmental conditions, such as underwater conditions. The abrasive wear resistance of coatings manufactured by a wet welding method was investigated and compared with those produced in air. In general, the results showed that the former method could be successfully used for marine and offshore structures repair in underwater conditions without the necessity of dragging those elements to the surface of the water [4]. The study of [5] demonstrated the high significance of the surface analysis methods used to study corroded coatings formed on the iron anchors unearthed from the Gudu ruins for the restoration of cultural relics. Based on the research a strategy to repair and to prevent the iron anchors from further corrosive degradation in the humid exhibition environment was proposed.

The plasma electrolytic oxidation (PEO) is an electrochemical and environmentally friendly method to generate thick and dense oxide coatings on metals. The properties of coatings can be easily tailored by changing processing parameters, such as polarization voltage or anodization time in a wide range of electrolytes. Two comprehensive reviews concerning the growth mechanism, microstructure, chemical composition, and various functional and application aspects of ceramic coatings prepared on zirconium (Zr) and Zr alloy substrates by the PEO technique are presented in this Special Issue [6,7]. Due to their unique properties, including remarkable biocompatibility, bioactivity, corrosion and wear resistance, and photoluminescence performance, ceramic coatings can be applied in medicine (dental or percutaneous implants), photocatalysis (water splitting), or nuclear industries (fuel tubes and coolant channel material). The excellent photoluminescence (PL) properties of the Ce^{3+}/Eu^{2+} doped Al_2O_3 coating produced by the PEO of aluminum were demonstrated in the study of [8]. The PL could be modulated by changing the concentration of Eu_2O_3 and CeO_2 particles in the electrolyte which, in turn, induced changes in the $Ce^{3+} \rightarrow Eu^{2+}$ energy transfer.

The anodization of aluminum results in a formation of porous anodic alumina (AAO), which is a well-known template used in various fields of nanotechnology. The AAO was used as a porous ceramic support for ultra-thin Zn-S layers to study their structural and chemical composition by standard infrared (IR) spectroscopy [9]. Due to the well-developed surface of the AAO, it was possible to precisely determine the structure of a Zn-S layer as thin as 20 nm, which was "invisible" to IR spectroscopy when deposited on a flat Si substrate. The study of the Al dissolution rate in various acids and solutions in the absence of anodic polarization sheds new light on the processes occurring during the nucleation and growth of AAO [10].

The guest editor would like to express her gratitude to all authors who have contributed to this Special Issue. Many thanks are also due to the reviewers for a constructive evaluation of the manuscripts and to the publisher for kind and efficient cooperation. It is my pleasure to introduce the collected articles for all potential readers who are interested in better understanding these selected phenomena induced by metal, ceramic, and metal–ceramic coatings produced by a broad range of technologies.

Funding: This research received no external funding.

Institutional Review Board Statement: Not applicable.

Informed Consent Statement: Not applicable.

Data Availability Statement: The data are available from the corresponding author upon request.

Conflicts of Interest: The author declares no conflict of interest.

References

1. Hong, S.; Wu, Y.; Wang, B.; Lin, J. Improvement in tribological properties of Cr12MoV cold work die steel by HVOF sprayed WC-CoCr cermet coatings. *Coatings* **2019**, *9*, 825. [CrossRef]
2. Kopec, M.; Kukla, D.; Yuan, X.; Rejmer, W.; Kowalewski, Z.L.; Senderowski, C. Aluminide thermal barrier coating for high temperature performance of MAR 247 nickel based superalloy. *Coatings* **2021**, *11*, 48. [CrossRef]
3. Lu, X.; Wang, S.; Xiong, T.; Wen, D.; Wang, G.; Du, H. Study on corrosion resistance and wear resistance of Zn-Al-Mg/ZnO composite coating prepared by cold spraying. *Coatings* **2019**, *9*, 505. [CrossRef]
4. Tomków, J.; Czupryński, A.; Fydrych, D. The abrasive wear resistance of coatings manufacture on high-strength low-alloy (HSLA) offshore steel in wet welding conditions. *Coatings* **2020**, *10*, 219. [CrossRef]
5. Mai, B.; Chen, Y.; Zhang, Y.; Huang, Y.; Wang, J.; Li, Y.; Cao, M.; Cao, J. Analysis of iron anchor diseases unearthed from Gudu ruins in Xianyang City, Shaanxi Province, China. *Coatings* **2022**, *12*, 381. [CrossRef]
6. Attarzadeh, N.; Ramana, C.V. Plasma electrolytic oxidation ceramic coatings on zirconium (Zr) and Zr-alloys: Part I—Growth mechanisms, microstructure, and chemical composition. *Coatings* **2021**, *11*, 634. [CrossRef]
7. Attarzadeh, N.; Ramana, C.V. Plasma electrolytic oxidation ceramic coatings on zirconium (Zr) and Zr-alloys: Part II—Properties and applications. *Coatings* **2021**, *11*, 620. [CrossRef]
8. Stojadinović, S.; Ćirić, A. Ce^{3+}/Eu^{2+} doped Al_2O_3 coatings formed by plasma electrolytic oxidation of aluminum: Photoluminescence enhancement by $Ce^{3+} \rightarrow Eu^{2+}$ energy transfer. *Coatings* **2019**, *9*, 819. [CrossRef]
9. Włodarski, M.; Putkonen, M.; Norek, M. Infrared absorption study of Zn-S hybrid and ZnS ultrathin films deposited on porous AAO ceramic support. *Coatings* **2020**, *10*, 459. [CrossRef]
10. Poznyak, A.; Pligovka, A.; Turavets, U.; Norek, M. On-aluminum and barrier anodic oxide: Meeting the challenges of chemical dissolution rate in various acids and solutions. *Coatings* **2020**, *10*, 875. [CrossRef]

Article

Improvement in Tribological Properties of Cr12MoV Cold Work Die Steel by HVOF Sprayed WC-CoCr Cermet Coatings

Sheng Hong [1,*], Yuping Wu [1,*], Bo Wang [1] and Jinran Lin [2,3]

1   College of Mechanics and Materials, Hohai University, 8 Focheng West Road, Nanjing 211100, China; 181608010027@hhu.edu.cn
2   College of Engineering, Nanjing Agricultural University, 40 Dianjiangtai Road, Nanjing 210031, China; linjinran@njau.edu.cn
3   Jiangsu Jinxiang Transmission Equipment Co., Ltd., 1 Qinglonghu Road, Huaian 223001, China
*   Correspondence: hongsheng@hhu.edu.cn (S.H.); wuyuping@hhu.edu.cn (Y.W.)

Received: 29 October 2019; Accepted: 3 December 2019; Published: 4 December 2019

Abstract: The main objective of this study was to develop an efficient coating to increase the wear resistance of cold work die steel at different temperatures. The microstructures of high-velocity oxygen-fuel (HVOF)-sprayed WC-CoCr coatings were evaluated using scanning electron microscopy (SEM) and transmission electron microscopy (TEM). The effect of temperature on the tribological properties of the coatings and the reference Cr12MoV cold work die steel were both investigated by SEM, environmental scanning electron microscopy (ESEM), X-ray diffraction (XRD), and a pin-on-disk high-temperature tribometer. The coating exhibited a significantly lower wear rate and superior resistance against sliding wear as compared to the die steel at each test temperature, whereas no major differences in terms of the variation tendency of the friction coefficient as a function of temperature were observed in both the coatings and the die steels. These can be attributed to the presence of nanocrystalline grains and the fcc-Co phase in the coating. Moreover, the wear mechanisms of the coatings and the die steels were compared and discussed. The coating presented herein provided a competitive approach to improve the sliding wear performance of cold work die steel.

Keywords: sliding wear; cold work die steel; HVOF; WC-CoCr; cermet

---

1. Introduction

Friction, as one of the major sources of energy dissipation between the contact surfaces, is the main cause of wear and is also a common phenomenon in many industries such as rolling, packaging and mineral processing, where cold work die steel is widely used [1]. Most engineering components require the application of advanced materials that are resistant to wear in order to prevent or decrease material losses due to wear, to reduce the downtime of the equipment, and to increase efficiency and component quality. In order to meet these requirements, various metals [2], ceramics [3], and surface materials [4,5] with a unique combination of corrosion and wear resistance and high-temperature stability are being investigated and considered as effective methods to tailor the properties of engineering components [4–8]. In particular, surface technologies have been applied to a wide variety of materials without the necessity of using expensive and time-consuming heat treatments or alloying techniques, which can significantly change surface properties and may beneficially enhance the wear resistance of engineering materials [6,7]. As one of the most commonly-used surface coating technologies, thermal spraying has played a major role in the development of solutions for new technological problems as well as in the understanding of friction and wear mechanisms [8–10].

Ceramic–metal (i.e., cermet) is formed by at least a hard and a tough metallic binder phase to achieve specific properties for particular applications as a result of its exceptional toughness, high mechanical strength, corrosion and wear resistance [11]. For carbide cermet coatings, the chemical stability and oxidation resistance during the spraying process, which depend on chemical composition, microstructure, and surface condition of raw powders, should be improved when coming to the most demanding applications. Besides, the proper selection of thermal spraying techniques and the optimization of spraying process are also important steps [12–14]. High-velocity oxygen-fuel (HVOF) spraying is an option for these sorts of problems, as well as for the suppression or reduction of the decarburization and dissolution of carbides. In recent decades, considerable attempts have been devoted to improving the mechanical and wear properties of HVOF-sprayed WC-based cermet coatings through the proper choice of carbide size, binder phases and processing parameters [15–20]. Yang et al. [15] found that a higher wear rate is found with increasing carbide size for HVOF-sprayed WC-Co coatings. Meanwhile, Wesmann and Espallargas [16] investigated the tribological behaviors of HVOF-sprayed WC-CoCr coatings containing different carbide sizes and found that the effect of primary carbide size on friction and wear is marginal and somewhat inconsistent. Bolelli et al. [18] compared the dry sliding wear behaviors of HVOF-sprayed WC-(W,Cr)$_2$C–Ni and WC-CoCr coatings. It was found that the WC-(W,Cr)$_2$C–Ni coating exhibited a better wear resistance than the WC-CoCr coating, particularly at higher temperatures. Qiao et al. [20] reported that the hardest and toughest WC-Co coatings with superior tribological performance are obtained with a hot, neutral flame during the HVOF spraying process. Furthermore, the addition of Al into an HVOF-sprayed nanostructured WC-12Co coating is an effective method to enhance wear resistance [21]. However, till now, there is only limited works focus on sliding wear of HVOF-sprayed WC-based coatings at different temperatures [22,23]. Despite a high volume of published work, the high temperature wear properties of HVOF-sprayed WC-CoCr coatings have not been entirely understood.

In the past, the authors have demonstrated that HVOF-sprayed WC-CoCr coatings exhibit superior tribological properties as compared to the Cr12MoV cold work die steel at different loads, owing to the presence of a soft binder and a homogeneous dispersion of fine WC grains within the coating [24]. To our knowledge, there is still lack of systematic research on the relationship between detailed microstructures and sliding wear properties of HVOF-sprayed WC-CoCr coatings at different temperatures. The present study aimed to discover the effect of temperature on the sliding wear behavior of the WC-CoCr coating and to illustrate the dry sliding wear mechanisms at different temperatures in relation to the microstructures.

2. Experimental Procedure

Commercially available WC-CoCr cermet powder (Large Solar Thermal Spraying Material Co. Ltd, Chengdu, China) with a particle size of 15~45 μm was used for HVOF spraying in this study. The nominal composition of the cermet powder was, by wt.%: 4.0 Cr, 10.0 Co, 5.3 C, and 80.7 W. The WC-CoCr coating was deposited on Cr12MoV cold work die steel substrates with a thickness of approximately 200 μm by using an HVOF spray system (Praxair Tafa-JP8000, Danbury, CT, USA). Before spraying, the substrates were degreased with acetone in an ultrasonic bath, dried in hot air, and then adequately sandblasted with alumina (550 μm). The deposition of the coatings was conducted at the following conditions: a kerosene flow rate of 0.38 L·min^{-1} (standard liter per minute (SLPM)), oxygen flow rate of 897 L·min^{-1}, spray distance of 300 mm, argon carrier gas flow rate of 10.86 L·min^{-1}, powder feed rate of 50 g·min^{-1}, and spray gun speed of 280 mm·s^{-1}.

A scanning electron microscopes (SEM, Hitachi S-3400N, Tokyo, Japan) equipped with an energy dispersive spectroscope (EDS, EX250), alongside a high-resolution transmission electron microscope (HRTEM, JEOL JEM-2100F, Tokyo, Japan) were utilized to characterize the microstructures of the coating. Specimens for the TEM examination were prepared by grinding, polishing, mechanical dimpling, and argon ion milling to achieve electron transparency. The porosity of the coating was calculated by randomly using an image analyzer on 20 optical microscopy (OM, Olympus BX51M, Tokyo, Japan) images with a magnification of 500 from polished cross sections of the coating.

Dry sliding friction and wear tests were conducted on a commercially available, pin-on-disk high-temperature tribometer (Beilun MG-2000, Zhangjiakou, China) according to standard ASTM G99-05 [25]. Before the test, all coating and die steel specimens were grinded by 240#, 400#, 600#, 800#, 1000#, 1500# and 2000# SiC abrasive papers, polished by 2.5 and 0.5 μm diamond pastes, cleaned in an ultrasonic bath with acetone, and subsequently dried in hot air. The average surface roughness (R_a) values of the specimens were 0.02 μm. In the test, the upper pin of Al_2O_3 ball with a diameter of 6 mm as the friction counterpart was stationary, while the counterface disk with a diameter of 45 mm and a thickness of 7 mm was rotated. The mating materials were HVOF-sprayed WC-CoCr coatings and Cr12MoV cold work die steel. Dry sliding tests were performed in air at room temperature (RT), 200 °C and 500 °C. The relative humidity varied between 35% and 55%. A normal load of 50 N and a sliding velocity of 0.9 m·s^{-1} for 30 min were applied. The frictional moments were consistently recorded by a computer. After the test, the R_a values and the wear tracks of all coating and die steel specimens were measured by a profile and roughness tester (Taiming JB-4C, Shanghai, China). The wear rate was calculated by dividing the volume loss by load and sliding distance. The morphologies of worn surface of the coatings were examined by SEM and EDS. The surfaces of the samples inside the wear tracks were further investigated by X-ray diffraction (XRD, Bruker D8-Advanced, Karlsruhe, Germany) with Cu Kα radiation (λ = 1.54 Å) and step 0.02° operated at 40 kV and 40 mA in order to characterize the structure of the oxide scale that was formed at the various testing temperatures. The XRD patterns were taken with a beam spot size of 0.4 mm × 2 mm, ensuring the focus of XRD optics and considerable diffraction intensity on the wear tracks. The worn surfaces of the Cr12MoV cold work die steel were characterized by environmental scanning electron microscopy (ESEM, Philips XL30, Hillsboro, Holland). Tests were conducted at least thrice for each condition to ensure the repeatability and reliability of the reported data.

3. Results and Discussion

3.1. Microstructural Analysis of WC-CoCr Coatings

Our previous studies [26,27] have shown that conventional WC-CoCr coatings can be successfully synthesized by HVOF spraying technology and an optimal spray parameter can be obtained. The coating was mainly composed of WC, W_2C and an amorphous phase. Figure 1 shows the typical cross-sectional morphologies of the coating deposited on the AISI 1045 steel substrate. This is revealed from the overall view shown in Figure 1a, which shows that the coating possessed a dense microstructure and the interface of the coating and substrate was compact without obvious stripping. The thickness value was about 200 μm for the coating. As shown in Figure 1b, few internal defects such as pores and microcracks were observed. The coating had a low porosity value of 0.77%, resulting from high flame velocity and low flame temperature of the HVOF spraying process, which restrains decarburization [27]. There was a homogeneous distribution of tungsten carbide grains, as shown in the magnified image of the rectangular frame in Figure 1b. Additionally, inter-lamellar oxidation is indicated by yellow arrow in Figure 1c, which may have been due to the reaction between tungsten and oxygen that came from the spraying chamber [28,29].

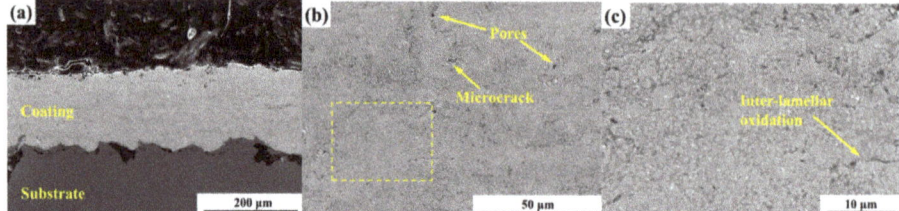

Figure 1. SEM images of a transverse section of the high-velocity oxygen-fuel (HVOF)-sprayed WC-CoCr coating: (**a**) an overall view morphology; (**b**) pores and microcracks; (**c**) a magnification of the rectangular frame in (**b**).

TEM and HRTEM images illustrating the detailed microstructure of the coating are shown in Figure 2. From Figure 2a, it can be seen that nanocrystalline grains were randomly oriented in the amorphous matrix and could be observed to be about 10 nm in size. The composition of amorphous matrix was Co–W–C, which was due to the high cooling rate of the splats. This is similar to phenomenon reported by Stewart et al. [30] and Li et al. [31]. According to the HRTEM image and selected area electron diffraction (SAED) pattern shown in Figure 2b,c, the nanocrystalline grains with an interplanar distance of 0.23 nm grew in parallel to the [100] zone axis, which corresponded to the W_2C phase. On the one hand, the size of nanocrystalline grains in the nanometer range was due to the presence of the amorphous phase, which can hinder grain growth during the coating process [32]. On the other hand, the formation of the nanocrystalline grains in the coating may lead to the enhancement of mechanical properties as a result of dispersion strengthening. A similar result was also observed by Lee et al. [33]. It is possible that the nanocrystalline/amorphous structure formed as a result of the solid state transformation and partial crystallization of the amorphous structure induced by the preferred oxidation on the in-flight particles surface and the thermal fluctuation from the subsequent splats [34,35]. The TEM analysis performed on the coating also indicated the presence of the fcc-Co phase, as shown in Figure 2d. The corresponding SAED pattern is shown in Figure 2e, which indicates that the Co phase was indexed to a face-centered cubic structure with the incident beam parallel to the zone axis of [320]. Human et al. [36] reported that compared to the hcp-Co phase, the fcc-Co phase is expected to be thermodynamically more stable. Hence, the presence of the fcc-Co phase is beneficial to the wear resistance of a coating at high temperatures.

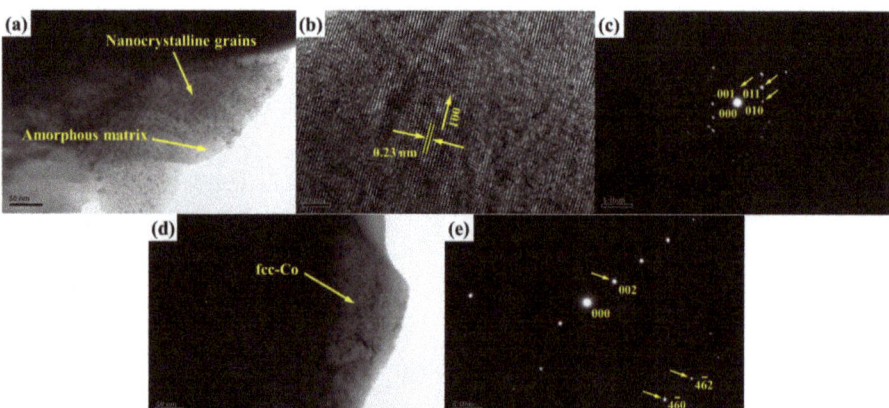

Figure 2. TEM images of typical microstructure of the HVOF-sprayed WC-CoCr coating: (**a**) a region of nanocrystalline grains and amorphous matrix; (**b**) high-resolution transmission electron microscopy (HRTEM) image of the nanocrystalline grain; (**c**) selected area electron diffraction (SAED) pattern of the W_2C phase; (**d**) a region of fcc-Co; and (**e**) SAED pattern of the fcc-Co phase.

3.2. Friction and Wear Properties

The friction coefficients data were continuously recorded by the high-temperature tribometer over a sliding time of 30 min. The typical friction coefficients of the HVOF-sprayed WC-CoCr coatings and Cr12MoV cold work die steel at different temperatures are shown in Figure 3a,b, respectively. It can be seen that all of the tests experienced a running-in period followed by a gradual stabilization in the friction coefficient value, which is consistent with the literature [37,38]. In all cases, the friction coefficient value was initially low and subsequently attained a steady state. This can be attributed to the evolution of tribo-oxidation products between the asperity contact of the tribopair at an early stage and the severe micro-fracture and subsequent pull-out of grains at a later stage of sliding [39]. Moreover, the fluctuation of the friction coefficient and the running-in period both decreased in duration with the increasing temperature, probably due to the lower initial resistance to the conformation between the two sliding surfaces as a result of the formation of larger amounts of the oxide films with lubricating properties at higher temperatures. This also suggests that there was a more efficient activation of wear protection processes at higher temperatures, which can be attributed to the more uniform distribution of the lubricating oxides over the wear track and the more active sites for oxidation induced by the larger fraction of interphase boundaries [40].

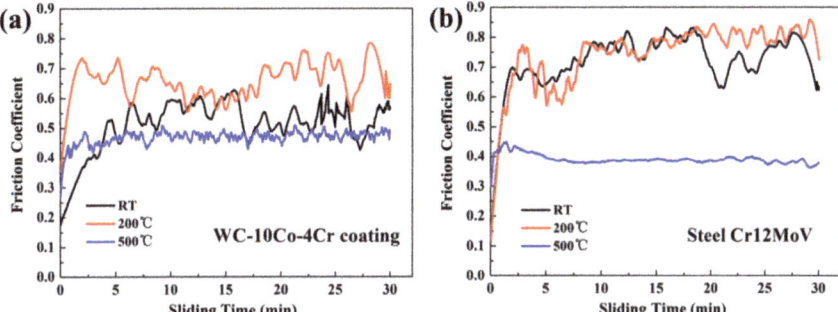

Figure 3. Friction coefficients varied with sliding time for HVOF-sprayed WC-CoCr coatings (**a**) and Cr12MoV cold work die steel (**b**) at different temperatures (counterface Al_2O_3, load 50 N, and sliding velocity 0.9 m·s^{-1}).

The average values for the friction coefficient during the steady period, R_a, of the worn surfaces and the wear rate are presented in Table 1. There was an approximately 25% increase in the friction coefficient of the coating when test temperature increased from RT to 200 °C. Additionally, it can be observed from Table 1 that the friction coefficient of the coating at 500 °C yielded the lowest friction coefficient value, which was about 29% lower than that at 200 °C. We can deduce that the wear track of the coating at 200 °C had a rougher contact surface and different oxide composition compared to that at RT, and the wear track of the coating at 500 °C appeared to be covered with friction oxide layers as a result of the oxidation of wear debris after the long-term action of high temperature contact between the coating and the Al_2O_3 ceramic. Furthermore, the interposition of the tribo-oxidation film between the coating surface and the Al_2O_3 ceramic ball at 500 °C prevented direct contact and decreased the friction coefficient. The findings in this study are similar to observations in a publication by Bolelli et al. [18], where the dry sliding wear behavior of HVOF-sprayed WC-(W,Cr)$_2$C–Ni and WC-CoCr hard metal coatings at different temperatures was studied. Wesmann and Espallargas [16] also reported that temperature has a greater effect on friction coefficient than the atmosphere and the carbide size of HVOF-sprayed WC-CoCr coatings. In our study, the average friction coefficients during the steady period of the Cr12MoV cold work die steel at different temperatures were 0.75, 0.79 and 0.38, respectively. It is worth noting that the variation tendency of friction coefficient as a function of temperature for the die steel showed a similar feature, although the reduction of the friction coefficient from 200 to 500 °C for the die steels was more significant compared to that for the coatings. In addition, the R_a value of worn surface and the wear rate both increased with increasing test temperature for the coatings and the die steels, which suggest that the sliding wear mechanism may be changed at different temperatures. For the coatings, the wear rate increased from 3.58×10^{-5} mm$^3 \cdot$N$^{-1} \cdot$m^{-1} at RT to 9.79×10^{-5} mm$^3 \cdot$N$^{-1} \cdot$m^{-1} at 500 °C. Similarly for the die steels, the wear rate firstly increased from 9.76×10^{-5} mm$^3 \cdot$N$^{-1} \cdot$m^{-1} at RT to 14.39×10^{-5} mm$^3 \cdot$N$^{-1} \cdot$m^{-1} at 200 °C, then climbed sharply to 150.4×10^{-5} mm$^3 \cdot$N$^{-1} \cdot$m^{-1} at 500 °C. These changes were mainly related to the content and distribution of oxides, as can be seen in detail in undermentioned worn surfaces analysis, as well as to the change of structures and mechanical properties [41]. The second reason for these changes may be that the hardness of the coating increased when test temperature increased from RT to 500 °C, whereas the fracture toughness showed the reverse tendency, resulting in the lowest wear resistance at 500 °C. This is similar to phenomenon reported by Wang et al. [42]. The third reason is the change in contact geometry, which resulted in a low friction coefficient. Moreover, the lubrication regime may have shifted during high temperature sliding, although the formation of tribo-oxidation films was more evident at higher temperature where they covered a larger fraction of the surface.

Table 1. Summary of friction and wear results for WC-10Co–4Cr coatings and Cr12MoV cold work die steel at different temperatures.

Characteristic Parameters	WC-10Co–4Cr Coatings			Cr12MoV Steel		
	RT	200 °C	500 °C	RT	200 °C	500 °C
Average friction coefficient during the steady period	0.53	0.66	0.47	0.75	0.79	0.38
Average surface roughness (R_a) values of worn surfaces, μm	0.13	0.16	0.21	0.28	0.45	2.27
Average wear rate, 10^{-5} mm$^3 \cdot$N$^{-1} \cdot$m^{-1}	3.58	6.29	9.79	9.76	14.39	150.4

In terms of the comparison between the HVOF-sprayed WC-CoCr coatings and Cr12MoV cold work die steel, the variation of the friction coefficients and wear rates as a function of temperature are shown in Figure 4a,b, respectively. It is clearly shown that the friction coefficients of the die steels were higher than those of the coatings at most test temperatures except at 500 °C (Figure 4a). The notable decrease of the friction coefficient from 200 to 500 °C for the die steels was caused by the absence of pores and oxide inclusions in steel and the formation of compact and smooth oxide films with a self-lubricant function on the surface. For each test temperature, the wear rates of the coatings were consistently lower than those of the die steels, especially at 500 °C (Figure 4b). This is because the coating with the fcc-Co phase and nanocrystalline grains could increase its stabilization, mechanical strength and resistance to plastic flow, therefore resulting in a lower wear rate of the coating at high temperatures. The die steel did not show an improved anti-wear performance at 500 °C, even though it presented the lowest friction coefficient, which can be attributed to the strong plastic deformation and local failure of the materials.

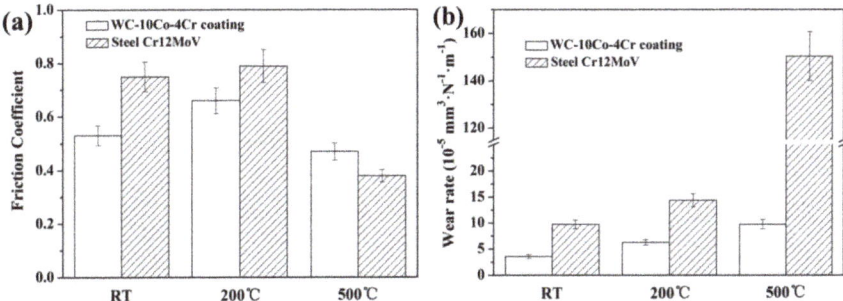

Figure 4. Variation of the friction coefficients (**a**) and wear rates (**b**) as a function of temperature for the HVOF-sprayed WC-CoCr coatings and Cr12MoV cold work die steel (counterface Al_2O_3, load 50 N, sliding velocity 0.9 m·s^{-1}).

3.3. Worn Surfaces Analysis

After dry sliding against Al_2O_3 ball at different temperatures, the worn surfaces of the HVOF-sprayed WC-CoCr coatings and Cr12MoV cold work die steel were investigated by SEM, ESEM and EDS in order to determine the predominant wear mechanism. Figure 5 presents the SEM images of the worn surfaces of the coatings after 30 min sliding with a load of 50 N and a sliding speed of 0.9 m·s^{-1}. An examination of the elemental content of non-rubbed surface and different regions of wear scars by the EDS analysis is shown in Table 2. From Figure 5a,b, it can be seen that the worn surface of the coating at RT exhibited relatively smooth characteristics with extrusion deformation sparsely distributed on it. Additionally, small amount of pits and carbide pull-out were detected, demonstrating that severe plastic deformation took place at the carbide–binder interface and cracks initiated at the defects under friction stress. Furthermore, Co, W and O were the primary elements according to the EDS analysis (wt.%) of point A, which suggests a possible formation of oxides. As the test temperature increased to 200 °C, severe wear traces with oxidized clusters could be seen spreading over the worn surface (Figure 5c) as could cracks, carbide pull-out, and adhesion between the coating and the Al_2O_3 ceramic (Figure 5d). The EDS results of point B illustrate a high weight percentage of W and O, confirming that a greater number of oxides were identified on the dark clusters as compared to that at RT. The EDS results of point C demonstrate that W, Al and O appeared on the white debris, although the quantities of the white debris were tiny. The existence of Al content confirms that the adhesive wear phenomenon was formed during the sliding process at 200 °C. The majority of the worn surface of the coating at 500 °C was covered by oxide wear debris and tribo-oxidation film (Figure 5e), which is verified by the EDS analysis of point D in Figure 5f and is consistent with previous studies [43]. The EDS spectrum of point D reveals that the dark grey materials primarily contained Co, W and O. During sliding wear at 500 °C, tribo-oxidation film may have broken up and led to the oxidation of the wear debris when it reached the critical thickness. In addition, wear tracks exhibited cracks and carbide pull-out, along with some regions of fatigue delamination. It is suggested that tribo-oxidation and fatigue wear contributed to the dominant wear mechanism of the coating at 500 °C, resulting in the higher wear rate. As shown in Figure 5 and Table 2, there were light gray areas (labeled as points E, F, and G) on worn surfaces at all test temperatures, and the regions in light gray tone mainly contained W and Co along with the presence of O. Compared with the non-rubbed surface, the content of O on the light gray areas was significantly increased, confirming that slight oxidation occurred along with wear track during sliding wear at all test temperatures, particularly at 500 °C. Moreover, the content of oxygen on the surface rubbed at 200 °C (positions B and C) was higher than those on the surface rubbed at 500 °C (positions D and G), which may have been due to the appearance of adhesive wear as a result of adhesion between the coating and the Al_2O_3 ceramic.

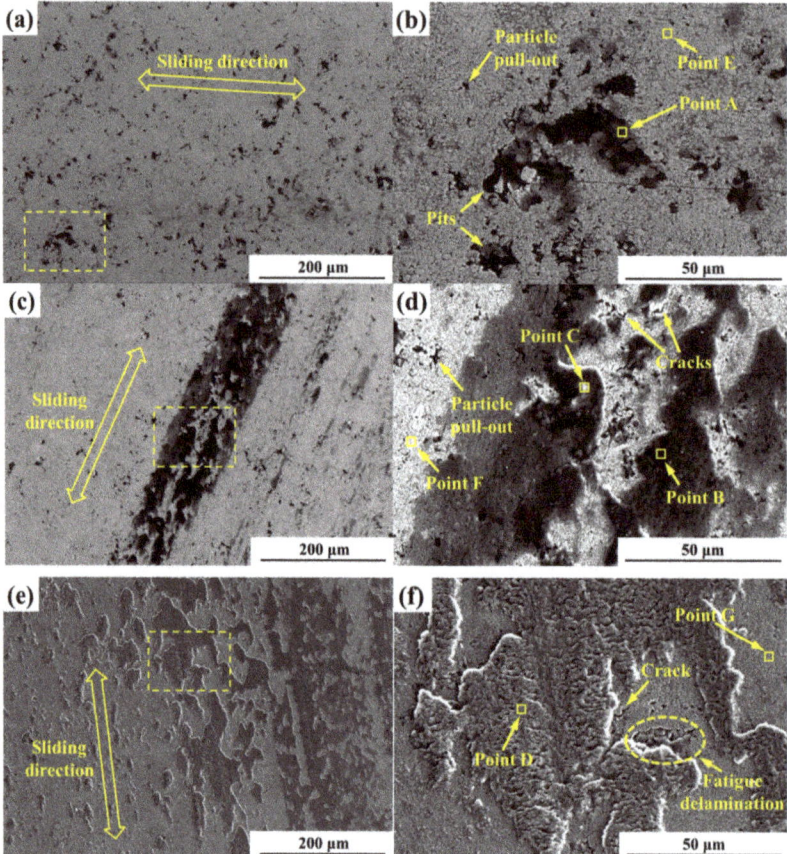

Figure 5. SEM images from worn surfaces of the HVOF-sprayed WC-CoCr coatings at different temperatures: (**a,b**) room temperature (RT); (**c,d**) 200 °C; and (**e,f**) 500 °C. (**b**) A magnification of the rectangular frame in (**a**). (**d**) A magnification of the rectangular frame in (**c**). (**f**) A magnification of the rectangular frame in (**e**).

Table 2. The energy dispersive spectroscopy (EDS) analysis results corresponding to the positions in Figure 5 and the non-rubbed surface.

Element, wt.%	Positions							
	A	B	C	D	E	F	G	Non-Rubbed
C K	1.74	–	1.69	–	10.79	10.10	3.49	21.55
O K	37.95	48.60	32.38	25.58	2.32	5.17	16.15	0.85
Al K	–	–	21.35	–	–	–	–	–
Cr K	2.46	2.99	2.24	2.85	1.59	0.93	2.28	5.30
Co K	8.23	7.37	6.33	8.72	11.05	9.67	11.34	13.96
W K	49.62	41.05	36.01	62.86	74.25	74.13	66.74	58.34

Figure 6 depicts representative ESEM images of worn surfaces of the die steels after 30 min sliding with load of 50 N and a sliding speed of 0.9 m·s^{-1}. As shown in Figure 6a, a large amount of wear debris in the shape of floccules could be observed on the wear track at RT. This can be explained by the fact that the microhardness of the die steel was 650 HV$_{0.1}$, well below that of the upper pin of Al$_2$O$_3$ ball. At the early stage of sliding, the die steel mainly experienced severe plastic deformation. In the progress of sliding wear, the forming, expanding and spalling of cracks took place under the stress of extrusion and friction, which led to the formation of wear debris and to the flocculent accumulation of debris. Figure 6b shows that shallow wear grooves, the spalling of carbides and strip-like carbides with the characters of clumped existed on the worn surface of the die steel at 200 °C. This shows that strip-like carbides are capable of withstanding wear grooves to some extent. The reason for this may be that the microhardness of the strip-like carbides (823 $HV_{0.1}$) was higher than that of the bulk-like carbides (747 $HV_{0.1}$), which is associated with the enhancement of strength and wear resistance in die steels. In addition, the Cr_7C_3-carbide, as evidenced by XRD analysis in Figure 7b, has a relatively high fracture toughness [44] that contributed to the positive effect on the wear resistance. Wayne et al. [45] found that the higher the fracture toughness, the lower the abrasive wear rate for a range of sintered WC-Co cermet. Huth et al. [46] also reported that M_7C_3-carbides can efficiently deflect wear path. Though the strip-like carbides potentially act as strengthening phases, their spalling is not negligible due to the irregular shape of carbides and the friction stress as the temperature increases, which causes severe abrasive wear and resulted in the growth of the wear rate, as seen in Figure 4b. As the test temperature increased to 500 °C (Figure 6c), a severe plowing phenomenon was observed on the worn surface of the die steel, and this contributed to the extremely high volume loss. Under the combined effect of internal stress, thermal stress, and friction stress, the wear debris and oxides formed furrows on the worn surface as a result of fragmentation and exfoliation, and then the abrasive wear obviously occurred.

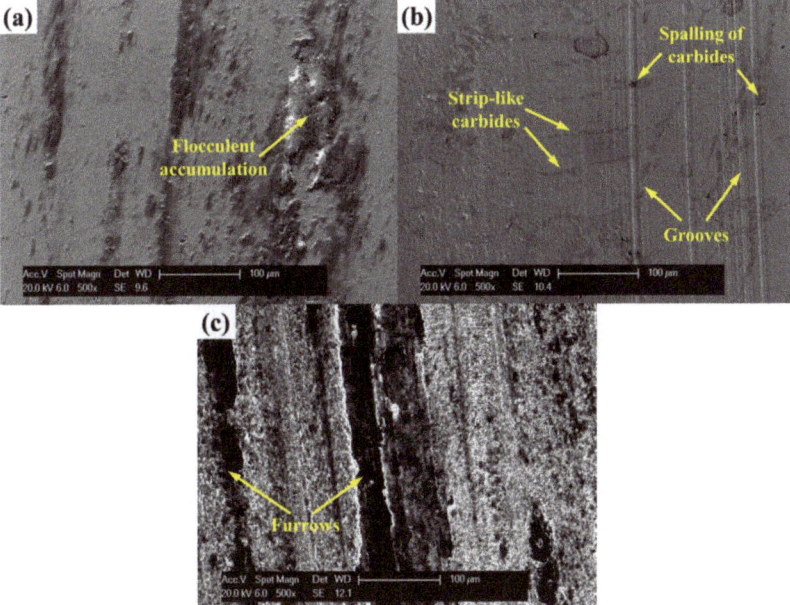

Figure 6. Environmental scanning electron microscopy (ESEM) images from the worn surfaces of Cr12MoV cold work die steel at different temperatures: (**a**) RT; (**b**) 200 °C; and (**c**) 500 °C.

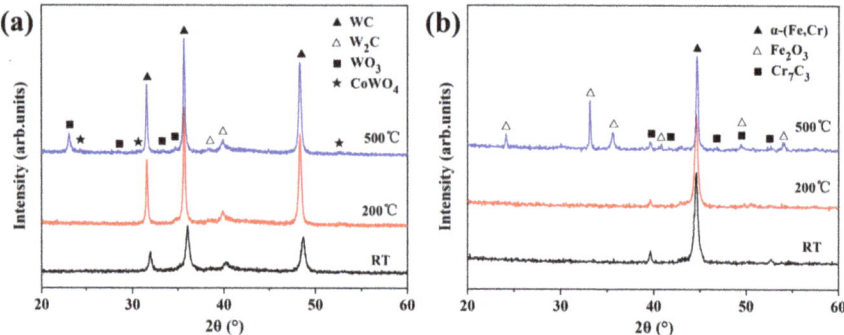

Figure 7. XRD patterns of the worn surface for the HVOF-sprayed WC-CoCr coatings (**a**) and Cr12MoV cold work die steel (**b**) at different temperatures (counterface Al_2O_3, load 50 N, and sliding velocity 0.9 m·s^{-1}).

The coating and die steel samples were analyzed with XRD to identify and understand the wear mechanisms responsible for the variation of microstructural constituents after the sliding at different temperatures. Figure 7 illustrates the XRD patterns of the worn surfaces of the HVOF-sprayed WC-CoCr coatings and Cr12MoV cold work die steel at different temperatures. The XRD analysis of the coating (Figure 7a) after sliding wear at RT and 200 °C identified the presence of WC and W_2C carbides, as expected when considering the coating composition. In the meanwhile, no evidence of the presence of oxides was found, which suggests that the content of oxides may have been too low to be detected by XRD analysis, and tribo-oxidation was not obvious at relatively low temperature. After sliding wear at 500 °C, WC and W_2C were the main phases, though some additional peaks corresponding to WO_3 alongside some minor $CoWO_4$ were also identified, suggesting a uniform distribution of oxide phases in the mixture. Comparing the XRD patterns of the coatings at RT and 200 °C, it could be seen that with the increasing test temperature, the intensities of oxide peaks increased (especially the peaks corresponding to WO_3). Similarly for the die steels (Figure 7b), a significant change in the XRD pattern at 500 °C was that new phase of Fe_2O_3 could be identified as compared to those at RT and 200 °C, which indicated the presence of a fraction of the oxide wear debris. Furthermore, the intensities of oxide peaks for the die steel were obviously higher than those for the coating after sliding wear at 500 °C. This may have been because of the presence of the fcc-Co phase (as evidenced in TEM analysis of the coating), whose oxidize rate is relatively low at high temperatures [36,47]. Therefore, the HVOF-sprayed WC-CoCr coating exhibits potential application on Cr12MoV cold work die steel.

4. Conclusions

The experiments were conducted by using HVOF spraying to fabricate a WC-CoCr cermet coating to improve the surface tribological properties of Cr12MoV cold work die steel. The main conclusions can be drawn from this work are as follows:

(1) The HVOF-sprayed WC-CoCr coating had a dense microstructure and compact interfacial bonding. A nanocrystalline/amorphous structure and an fcc-Co phase could be obtained in this coating, which could increase the mechanical strength, reduce the oxidize rate and maintain the resistance to sliding wear of the coating in comparison with die steel.

(2) Dry sliding friction and wear tests indicated that the R_a value of the worn surface and the wear rate both increased with increasing test temperature for the coatings and die steels, while the friction coefficient firstly increased as the test temperature increased from RT to 200 °C and then decreased from 200 to 500 °C. For each test temperature, the R_a value of worn surface and the wear rates of the coatings were consistently lower than those of the die steels, especially at 500 °C.

(3) After the sliding wear at RT and 200 °C for the coating, the presence of oxides was not identified by XRD analysis. The intensities of WO_3 and $CoWO_4$ peaks increased as the test temperature reached 500 °C. As for the die steels, a new phase of Fe_2O_3 was identified in the XRD pattern at 500 °C as compared to those at RT and 200 °C.

(4) The tribological properties of coated steel were improved, which could be attributed to its high hardness, low surface roughness, and friction coefficient resulting from the formation of nanocrystalline/amorphous structure and the fcc-Co phase in the coating.

(5) The sliding wear mechanisms of the coatings were extrusion deformation at RT, carbide pull-out and adhesive wear at 200 °C, and tribo-oxidation wear and fatigue wear at 500 °C. The mechanisms involved in the sliding wear process of the die steels with the increase of the temperature were plastic deformation, the flocculent accumulation of debris, the formation of strip-like carbides, furrows, and abrasive wear.

Author Contributions: Conceptualization, S.H. and Y.W.; Methodology, S.H. and B.W.; Investigation, S.H. and J.L.; Writing—Original Draft Preparation, S.H.; Supervision, S.H. and Y.W.

Funding: The research was supported by the National Natural Science Foundation of China (Grant Nos. 51979083 and 51609067), the Fundamental Research Funds for the Central Universities (Grant No. 2018B17014), the China Postdoctoral Science Foundation (Grant Nos. 2018T110435 and 2017M621665), the Postdoctoral Science Foundation of Jiangsu Province (Grant No. 2018K022A), and Shuangchuang Program of Jiangsu Province.

Conflicts of Interest: The authors declare no conflict of interest.

References

1. Holmberg, K.; Erdemir, A. Influence of tribology on global energy consumption, costs and emissions. *Friction* **2017**, *5*, 263–284. [CrossRef]
2. Grabon, W.; Pawlus, P.; Wos, S.; Koszela, W.; Wieczorowski, M. Effects of cylinder liner surface topography on friction and wear of liner-ring system at low temperature. *Tribol. Int.* **2018**, *121*, 148–160. [CrossRef]
3. Szala, M.; Hejwowski, T. Cavitation erosion resistance and wear mechanism model of flame-sprayed Al_2O_3–40%TiO_2/NiMoAl cermet coatings. *Coatings* **2018**, *8*, 254. [CrossRef]
4. Bayer, I.S. On the durability and wear resistance of transparent superhydrophobic coatings. *Coatings* **2017**, *7*, 12. [CrossRef]
5. Silva, F.; Martinho, R.; Andrade, M.; Baptista, A.; Alexandre, R. Improving the wear resistance of moulds for the injection of glass fibre-reinforced plastics using PVD coatings: A comparative study. *Coatings* **2017**, *7*, 28. [CrossRef]
6. Gachot, C.; Rosenkranz, A.; Hsu, S.M.; Costa, H.L. A critical assessment of surface texturing for friction and wear improvement. *Wear* **2017**, *372*, 21–41. [CrossRef]
7. Tian, L.H.; Feng, Z.K.; Xiong, W. Microstructure, microhardness, and wear resistance of AlCoCrFeNiTi/Ni60 coating by plasma spraying. *Coatings* **2018**, *8*, 112. [CrossRef]
8. Wang, Y.; Zhao, Y.L.; Darut, G.; Poirier, T.; Stella, J.; Wang, K.; Liao, H.L.; Planche, M.P. A novel structured suspension plasma sprayed YSZ-PTFE composite coating with tribological performance improvement. *Surf. Coat. Technol.* **2019**, *358*, 108–113. [CrossRef]
9. Singh, J.; Kumar, S.; Mohapatra, S.K. Tribological performance of Yttrium (III) and Zirconium (IV) ceramics reinforced WC-10Co4Cr cermet powder HVOF thermally sprayed on X2CrNiMo-17-12-2 steel. *Ceram. Int.* **2019**, *45*, 23126–23142. [CrossRef]
10. Singh, J.; Kumar, S.; Mohapatra, S.K. Tribological analysis of WC-10Co-4Cr and Ni-20Cr_2O_3 coating on stainless steel 304. *Wear* **2017**, *376*, 1105–1111. [CrossRef]
11. Zhang, Y.Q.; Hong, S.; Lin, J.R.; Zheng, Y. Influence of ultrasonic excitation sealing on the corrosion resistance of HVOF-sprayed nanostructured WC-CoCr coatings under different corrosive environments. *Coatings* **2019**, *9*, 724. [CrossRef]
12. Yuan, J.H.; Ma, C.W.; Yang, S.L.; Yu, Z.S.; Li, H. Improving the wear resistance of HVOF sprayed WC-Co coatings by adding submicron-sized WC particles at the splats' interfaces. *Surf. Coat. Technol.* **2016**, *285*, 17–23. [CrossRef]

13. Geng, Z.; Hou, S.H.; Shi, G.L.; Duan, D.L.; Li, S. Tribological behaviour at various temperatures of WC-Co coatings prepared using different thermal spraying techniques. *Tribol. Int.* **2016**, *104*, 36–44. [CrossRef]
14. Da Silva, F.S.; Cinca, N.; Dosta, S.; Cano, I.G.; Couto, M.; Guilemany, J.M.; Benedetti, A.V. Corrosion behavior of WC-Co coatings deposited by cold gas spray onto AA 7075-T6. *Corros. Sci.* **2018**, *136*, 231–243. [CrossRef]
15. Yang, Q.Q.; Senda, T.; Ohmori, A. Effect of carbide grain size on microstructure and sliding wear behavior of HVOF-sprayed WC–12% Co coatings. *Wear* **2003**, *254*, 23–34. [CrossRef]
16. Wesmann, J.A.R.; Espallargas, N. Effect of atmosphere, temperature and carbide size on the sliding friction of self-mated HVOF WC-CoCr contacts. *Tribol. Int.* **2016**, *101*, 301–313. [CrossRef]
17. Mateen, A.; Saha, G.C.; Khan, T.I.; Khalid, F.A. Tribological behaviour of HVOF sprayed near-nanostructured and microstructured WC-17wt.%Co coatings. *Surf. Coat. Technol.* **2011**, *206*, 1077–1084. [CrossRef]
18. Bolelli, G.; Berger, L.M.; Bonetti, M.; Lusvarghi, L. Comparative study of the dry sliding wear behaviour of HVOF-sprayed WC-(W,Cr)$_2$C-Ni and WC-CoCr hardmetal coatings. *Wear* **2014**, *309*, 96–111. [CrossRef]
19. Hong, S.; Wu, Y.P.; Wang, B.; Zheng, Y.G.; Gao, W.W.; Li, G.Y. High-velocity oxygen-fuel spray parameter optimization of nanostructured WC-10Co-4Cr coatings and sliding wear behavior of the optimized coating. *Mater. Des.* **2014**, *55*, 286–291. [CrossRef]
20. Qiao, Y.F.; Fischer, T.E.; Dent, A. The effects of fuel chemistry and feedstock powder structure on the mechanical and tribological properties of HVOF thermal-sprayed WC-Co coatings with very fine structures. *Surf. Coat. Technol.* **2003**, *172*, 24–41. [CrossRef]
21. Basak, A.K.; Celis, J.P.; Vardavoulias, M.; Matteazzi, P. Effect of nanostructuring and Al alloying on friction and wear behaviour of thermal sprayed WC-Co coatings. *Surf. Coat. Technol.* **2012**, *206*, 3508–3516. [CrossRef]
22. Rovatti, L.; Lecis, N.; Dellasega, D.; Russo, V.; Gariboldi, E. Influence of aging in the temperature range 250–350 °C on the tribological performance of a WC-CoCr coating produced by HVOF. *Int. J. Refract. Met. Hard Mater.* **2018**, *75*, 218–224. [CrossRef]
23. Hong, S.; Wu, Y.P.; Wang, B.; Zhang, J.F.; Zheng, Y.; Qiao, L. The effect of temperature on the dry sliding wear behavior of HVOF sprayed nanostructured WC-CoCr coatings. *Ceram. Int.* **2017**, *43*, 458–462. [CrossRef]
24. Wu, Y.P.; Wang, B.; Hong, S.; Zhang, J.F.; Qin, Y.J.; Li, G.Y. Dry sliding wear properties of HVOF sprayed WC-10Co-4Cr coating. *Trans. Indian Inst. Metals* **2015**, *68*, 581–586. [CrossRef]
25. *ASTM G99-05 Standard Test Method for Wear Testing with a Pin-on-Disk Apparatus*; ASTM International: West Conshohocken, PA, USA, 2010.
26. Hong, S.; Wu, Y.P.; Gao, W.W.; Wang, B.; Guo, W.M.; Lin, J.R. Microstructural characterisation and microhardness distribution of HVOF sprayed WC-10Co-4Cr coating. *Surf. Eng.* **2014**, *30*, 53–58. [CrossRef]
27. Hong, S.; Wu, Y.P.; Gao, W.W.; Zhang, J.F.; Zheng, Y.G.; Zheng, Y. Slurry erosion-corrosion resistance and microbial corrosion electrochemical characteristics of HVOF sprayed WC-10Co-4Cr coating for offshore hydraulic machinery. *Int. J. Refract. Met. Hard Mater.* **2018**, *74*, 7–13. [CrossRef]
28. Schwetzke, R.; Kreye, H. Microstructure and properties of tungsten carbide coatings sprayed with various High-velocity oxygen fuel spray systems. *J. Therm. Spray Technol.* **1999**, *8*, 433–439. [CrossRef]
29. Niu, Y.R.; Zheng, X.B.; Ji, H.; Qi, L.J.; Ding, C.X.; Chen, J.L.; Luo, G.N. Microstructure and thermal property of tungsten coatings prepared by vacuum plasma spraying technology. *Fusion Eng. Des.* **2010**, *85*, 1521–1526. [CrossRef]
30. Stewart, D.A.; Shipway, P.H.; McCartney, D.G. Microstructural evolution in thermally sprayed WC-Co coatings: Comparison between nanocomposite and conventional starting powders. *Acta Mater.* **2000**, *48*, 1593–1604. [CrossRef]
31. Li, C.J.; Ohmori, A.; Harada, Y. Formation of an amorphous phase in thermally sprayed WC-Co. *J. Therm. Spray Technol.* **1996**, *5*, 69–73. [CrossRef]
32. Jiang, N.; Shen, Y.G.; Mai, Y.W.; Chan, T.; Tung, S.C. Nanocomposite Ti-Si-N films deposited by reactive unbalanced magnetron sputtering at room temperature. *Mater. Sci. Eng. B* **2004**, *106*, 163–171. [CrossRef]
33. Lee, K.H.; Chang, D.; Kwon, S.C. Properties of electrodeposited nanocrystalline Ni-B alloy films. *Electrochim. Acta* **2005**, *50*, 4538–4543. [CrossRef]
34. Sobolev, V.V.; Guilemany, J.M.; Miguel, J.R.; Calero, J.A. Investigation of the development of coating structure during high velocity oxy-fuel (HVOF) spraying of WC-Ni powder particles. *Surf. Coat. Technol.* **1996**, *82*, 114–120. [CrossRef]

35. Sharma, P.; Majumdar, J.D. Surface characterization and mechanical properties evaluation of Boride-dispersed Nickel-based coatings deposited on copper through thermal spray routes. *J. Therm. Spray Technol.* **2012**, *21*, 800–809. [CrossRef]
36. Human, A.M.; Roebuck, B.; Exner, H.E. Electrochemical polarisation and corrosion behaviour of cobalt and Co (W, C) alloys in 1 N sulphuric acid. *Mater. Sci. Eng. A* **1998**, *241*, 202–210. [CrossRef]
37. Blau, P.J. *Friction and Wear Transitions of Materials*; Noyes Publishing: Park Ridge, NJ, USA, 1989.
38. Zhu, Y.C.; Yukimura, K.; Ding, C.X.; Zhang, P.Y. Tribological properties of nanostructured and conventional WC-Co coatings deposited by plasma spraying. *Thin Solid Films* **2001**, *388*, 277–282. [CrossRef]
39. Wu, L.; Chen, J.X.; Liu, M.Y.; Bao, Y.W.; Zhou, Y.C. Reciprocating friction and wear behavior of Ti_3AlC_2 and Ti_3AlC_2/Al_2O_3 composites against AISI52100 bearing steel. *Wear* **2009**, *266*, 158–166. [CrossRef]
40. Lekatou, A.; Sioulas, D.; Karantzalis, A.E.; Grimanelis, D. A comparative study on the microstructure and surface property evaluation of coatings produced from nanostructured and conventional WC-Co powders HVOF-sprayed on Al7075. *Surf. Coat. Technol.* **2015**, *276*, 539–556. [CrossRef]
41. Shipway, P.H.; McCartney, D.G.; Sudaprasert, T. Sliding wear behaviour of conventional and nanostructured HVOF sprayed WC-Co coatings. *Wear* **2005**, *259*, 820–827. [CrossRef]
42. Wang, Q.; Li, L.X.; Yang, G.B.; Zhao, X.Q.; Ding, Z.X. Influence of heat treatment on the microstructure and performance of high-velocity oxy-fuel sprayed WC–12Co coatings. *Surf. Coat. Technol.* **2012**, *206*, 4000–4010. [CrossRef]
43. Geng, Z.; Li, S.; Duan, D.L.; Liu, Y. Wear behavior of WC-Co HVOF coatings at different temperatures in air and argon. *Wear* **2015**, *330*, 348–353. [CrossRef]
44. Guo, J.; Ai, L.Q.; Wang, T.T.; Feng, Y.L.; Wan, D.C.; Yang, Q.X. Microstructure evolution and micro-mechanical behavior of secondary carbides at grain boundary in a Fe-Cr-W-Mo-V-C alloy. *Mater. Sci. Eng. A* **2018**, *715*, 359–369. [CrossRef]
45. Wayne, S.F.; Baldoni, J.G.; Buljan, S.T. Abrasion and erosion of WC-Co with controlled microstructures. *Tribol. Trans.* **1990**, *33*, 611–617. [CrossRef]
46. Huth, S.; Krasokha, N.; Theisen, W. Development of wear and corrosion resistant cold-work tool steels produced by diffusion alloying. *Wear* **2009**, *267*, 449–457. [CrossRef]
47. Zhang, Y.Q.; Li, C.Y.; Xu, Y.H.; Tang, Q.H.; Zheng, Y.; Liu, H.W.; Fernandez-Rodriguez, E. Study on propellers distribution and flow field in the oxidation ditch based on two-phase CFD model. *Water* **2019**, *11*, 2506. [CrossRef]

© 2019 by the authors. Licensee MDPI, Basel, Switzerland. This article is an open access article distributed under the terms and conditions of the Creative Commons Attribution (CC BY) license (http://creativecommons.org/licenses/by/4.0/).

Communication

Aluminide Thermal Barrier Coating for High Temperature Performance of MAR 247 Nickel Based Superalloy

Mateusz Kopec [1,2,*], Dominik Kukla [1], Xin Yuan [2], Wojciech Rejmer [3], Zbigniew L. Kowalewski [1] and Cezary Senderowski [3]

[1] Institute of Fundamental Technological Research, Polish Academy of Sciences, Pawińskiego 5B, 02-106 Warszawa, Poland; dkukla@ippt.pan.pl (D.K.); zkowalew@ippt.pan.pl (Z.L.K.)
[2] Department of Mechanical Engineering, Imperial College London, London SW7 2AZ, UK; xin.yuan15@imperial.ac.uk
[3] Department of Materials Technology and Machinery, University of Warmia and Mazury, Oczapowskiego 11 St., 10-719 Olsztyn, Poland; wojciech.rejmer@uwm.edu.pl (W.R.); cezary.senderowski@uwm.edu.pl (C.S.)
* Correspondence: mkopec@ippt.pan.pl

Abstract: In this paper, mechanical properties of the as-received and aluminide layer coated MAR 247 nickel based superalloy were examined through creep and fatigue tests. The aluminide layer of 20 μm was obtained through the chemical vapor deposition (CVD) process in the hydrogen protective atmosphere for 8 h at the temperature of 1040 °C and internal pressure of 150 mbar. A microstructure of the layer was characterized using the scanning electron microscopy (SEM) and X-ray Energy Dispersive Spectroscopy (EDS). It was found that aluminide coating improve the high temperature fatigue performance of MAR247 nickel based superalloy at 900 °C significantly. The coated MAR 247 nickel based superalloy was characterized by the stress amplitude response ranging from 350 MPa to 520 MPa, which is twice as large as that for the uncoated alloy.

Keywords: chemical vapor deposition; nickel alloys; aluminide coatings; high temperature fatigue; creep

1. Introduction

Nickel alloys are characterized by their superior, high temperature performance properties including corrosion, heat and creep resistance. Hence, they are commonly used in aircraft engines [1,2]. Higher engine combustion temperatures tend to improve the efficiency of the propulsion systems, however, extremely high engine temperature requires an advanced materials that could withstand the operational loads in such aggressive environment [3–5]. The most conventional superalloys for gas turbines are MAR 247 [6], Rene 80 [7] and IN738 [8]. MAR 247 nickel based superalloy exhibit higher strength properties in comparison to these alloys, especially at high temperature. Its yield strength at 1000 °C is approximately 100 MPa greater that for IN738 and almost two times greater than for Rene 80 [6–8]. Additionally, MAR 247 nickel based superalloy was characterized by improved creep response in comparison to the mentioned alloys. The development of casting technologies led to the fabrication of new generation, cost effective alloys doped with Re and Ru to further increase the temperature stability [9]. However, these elements are expensive and their supply is strategically dangerous [9].

Among the promising solutions that could enhance the high temperature performance of engine elements, one can indicate the application of Ni-Al type intermetallics as coatings material. In such materials, due to their crystalline structure with strong chemical bonds and tightly packed atoms in the lattice, a relatively low diffusion of atoms during recrystallisation occurs. Thus, they are more resistant to creep and high-temperature corrosion in comparison to the classical superalloys. Among the NiAl-type intermetallics, only two phases (NiAl and Ni$_3$Al) would allow to form the coating structure, that could transfer high mechanical loads in aggressive corrosive and erosive environments, also at high temperature [10,11]. Ni$_3$Al phase based intermetallics, exhibit improved fatigue strength at high temperature in comparison to the commonly used Ni-based superalloys [12]. On the other hand, the ordered NiAl phase is characterized by a higher melting point (1640 °C), lower density (5.86 g/cm^3), higher oxidation resistance and higher thermal conductivity (λ = 76 W m^{-1} K^{-1}) in comparison to Ni$_3$AL (7.5 g/cm^3, λ = 21.4 W m^{-1} K^{-1}) [13,14]. Hence, the superior properties of NiAl and Ni$_3$Al aluminides at high temperature could be potentially utilize as protective coatings for aircraft engine turbines. Generally, thermal barrier coatings are applied to parts of gas turbine to reduce their operational temperature by approximately 100–300 °C and simultaneously increase their service life [15]. The coatings protect external surfaces of the material when it is in contact with hot gas atmosphere. The additional content of aluminium and chromium in coating allows to form a thermodynamically stable oxide layer that acts as diffusion barrier [16]. Such barrier protects the substrate material and significantly reduce its chemical reactivity during performance in aggressive environment [17–19]. The effect of protective coatings on high temperature performance of nickel based alloys was widely studied in literature. Duplex chromium/aluminium coatings were used to improve the creep performance over 30 years ago [20,21]. It was found that the aluminide coatings deposited on nickel superalloys could enhance their mechanical properties at elevated temperature [22,23] through significant reduction of oxidation and carbonisation during high temperature performance [24–27]. Functional properties of aluminide layer deposited on the nickel based superalloys in CVD process have been studied by Yaworska et al. [28]. It should be mentioned that, aluminide thermal barrier coatings for nickel superalloys were found useful in maintaining their superior mechanical properties, especially at high temperature. It has been also found that aluminium-based materials for aluminide coatings enhance the oxidation resistance [29,30] and high temperature performance of nickel superalloys without deterioration of its mechanical properties [31–33]. Since high temperature oxidation contributes to the reduction of the fatigue life at elevated temperatures [19], the application of thermal barrier coatings was found to be extremely important in terms of materials fatigue life. The general comparison of some basic mechanical properties of coatings and nickel based materials for turbine blades at high temperature were recently presented in comprehensive review by Wee et al. [9]. However, the high temperature mechanical response of coated MAR 247 nickel based superalloy were not studied as yet.

Therefore, the main aim of this work was to assess the high temperature mechanical response of conventional MAR 247 nickel based superalloy protected by aluminide coating (TBC) deposited by optimized chemical vapor deposition (CVD) process [34] through creep and fatigue tests. Such tests enabled both, characterization of the mechanical properties of material in question and assessment of an actual effectiveness of the CVD process parameters optimized in authors previous paper [34].

2. Materials and Methods

MAR 247 nickel based superalloy specimens were manufactured using casting process and uniform crystallization performed in ceramic moulds. Specimen casting was carried out in an ALD vacuum furnace. Specimens made of MAR 247 nickel based superalloy with directional grain orientation (DS) were transferred outside of the furnace under controlled speed of 3 mm/min. Specimens with equiaxed microstructure (EQ) were quenched in the furnace to achieve the required microstructure. The chemical composition of MAR247 nickel based superalloy was presented in Table 1.

Table 1. Chemical composition of MAR 247 nickel based superalloy (wt.%).

C	Cr	Mn	Si	W	Co	Al	Ni
0.09	8.80	0.10	0.25	9.70	9.50	5.70	bal.

Aluminide coatings were produced by chemical vapour deposition (CVD) using Ion-Bond setup (Ion Bond Bernex BPX Pro 325 S, IHI Ion bond AG, Olten, Switzerland) located in the Materials Testing Laboratory for the Aviation Industry of the Rzeszów

University of Technology, Rzeszów, Poland. It was a low activity aluminium process. The optimised CVD processes were executed in the hydrogen protective atmosphere for 8 h at the temperature of 1040 °C and internal pressure of 150 mbar [34]. The microstructural characterization and chemical composition analysis of the coatings were examined using Hitachi 2600N scanning electron microscope with Energy Dispersive Spectroscopy (EDS) attachment (Oxford Instruments, Oxford, UK) and HITACHI 260 (Hitachi, Tokyo, Japan) also with an EDS detector. The microhardness of the as-received and coated material was determined on a ZWICK hardness tester (Materialprüfung 3212002, Ulm, Germany) using the Vickers method. Standard and high temperature fatigue tests were performed on the MTS 810 testing machine (MTS System, Eden Prairie, MN, USA) equipped with the conventional and high temperature MTS extensometers. Uniaxial tensile tests were performed at strain rate equal to 2×10^{-4} s^{-1}. Fatigue tests at room and high temperatures were force controlled under zero mean value and constant stress amplitude with a frequency of 20 Hz. The range of stress amplitude from 350 MPa to 650 MPa was established on the basis of the yield point $R_{0.2}$ determined from the uniaxial tensile test. High temperature fatigue tests were carried out at 900 °C. The geometry of specimens is presented in Figure 1a. Creep tests at 600 °C under constant stress levels within the range from 700 MPa to 780 MPa were conducted on the standard creep testing machines. The strain measured by two independent sensors were recorded in time internals of 5 min, by two independent sensors. Engineering drawing of the creep test specimens is shown in Figure 1b. The Young's modulus was determined by means of the non-destructive tests in Department of Materials Technology, University of Warmia and Mazury using RFDA (Resonance Frequency and Damping Analyzer, Integrated Material Control Engineering NV, Genk, Belgium) according rules elaborated in the Gordon Laboratory, University of Cambridge [35] and the ASTME 1876-99 standard [36] and this method, an automatic impulser was applied to excite flexural vibrations in the specimen by striking lightly at the center of the specimen. Subsequently, a contactless microphone placed near the end of the specimen was used to register the specimen's mechanical response. Determination of the resonant frequency of natural vibrations damped by internal friction enabled evaluation of the Young's modulus under plane strain state conditions in the elastic range using empirical equations from the Euler-Bernoulli theory [34]. The Young's modulus measurements were carried out at the temperature range from 26 °C to 900 °C, during annealing and cooling of the specimen at the rate of 3 °C/min with subsequent annealing at target temperature for 30 min.

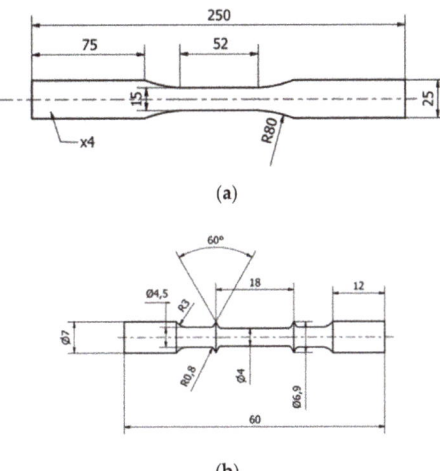

Figure 1. Geometry of the specimen used for: (**a**) uniaxial, standard and high temperature fatigue tests; (**b**) creep tests.

3. Results and Discussion

3.1. Microstructural Characterization of Coating during CVD Process

The structure of the NiAl coating was determined by the growth kinetics of the formed layers and conditioned by the temperature, pressure and synthesis time in the process of chemical vapor deposition. The general view of coating obtained was presented in Figure 2a. It was characterized by uniform thickness of 20 (±5 μm) and was evenly distributed on the MAR 247 nickel based superalloy surface. The intermetallic coating exhibits two-layer structure with approx. 11 μm thick homogeneous zone of secondary solid solution of the β (NiAl) phase and approx. 12 μm heterogeneous NiAl matrix with Ni_3Al phase dispersions found within its structure. The interlayer zone was characterized by the lower content of aluminium and the participation of Co, Cr and Ti alloying elements diffusing from the core of raw material (MAR 247 nickel based superalloy) under specific CVD process conditions. The chemical composition was analysed in the cross-section of specimens at 7 points starting from the edge into the material core (Figure 2b, Table 2). The highest content of aluminium (~23%) was found near the surface. The mass volume of aluminium did not change significantly in the cross-section of specimen (x1–x3) as its content of ~20% was also observed in lower layers (x4–x6). Additionally, the slight increase of chromium and cobalt can be observed in the central part of layer. The surface of the layer consists of large, NiAl intermetallic crystallites that formed an intermetallic superstructure (secondary solid solution β with B2 ordered structure) [34]. One can indicate that CVD conditions applied allowed to obtain a good coating coherence to the substrate since no defects were observed between coated material and coating itself. Some discontinuities found between sub-layers were caused by the extensive grinding during preparation of the specimens for the microstructural characterization.

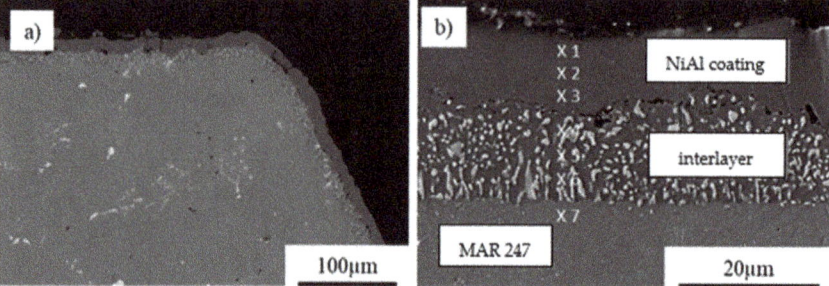

Figure 2. Microstructure of intermetallic coating produced by CVD method on MAR 247 nickel based superalloy at 1040 °C: (a) general view of the coating; (b) view of coating with points of EDS analysis marked.

Table 2. Chemical composition (wt.%) of coating surface obtained after deposition at 1040 °C.

Point	Al	Si	Ti	Cr	Co	Ni
x1	22.99	-	-	2.91	8.37	65.33
x2	21.92	-	0.40	4.30	9.13	64.25
x3	20.63	-	0.69	5.35	9.69	63.65
x4	19.59	0.45	0.86	5.78	10.05	63.28
x5	18.49	0.66	1.02	6.34	10.50	63.00
x6	18.25	0.69	1.04	6.68	9.87	63.48
x7	7.24	1.98	0.94	6.33	9.11	66.13

The hardness distribution was measured from the edge of the coating in its cross-section, (Figure 3). The microhardness of the near-surface NiAl coating zone was equal to 664 ± 25 HV0.05 and decreases towards the surface layer of MAR 247 nickel based superalloy to 450 ± 20 HV0.05. The hardness of transition area was approx. 550 ± 40 HV0.05.

The fluctuations founded in this area were related to the nonhomogeneous structure consisted of the hard NiAl phase and softer Ni$_3$Al phase. It was observed that hardness varied with the content of different alloying elements.

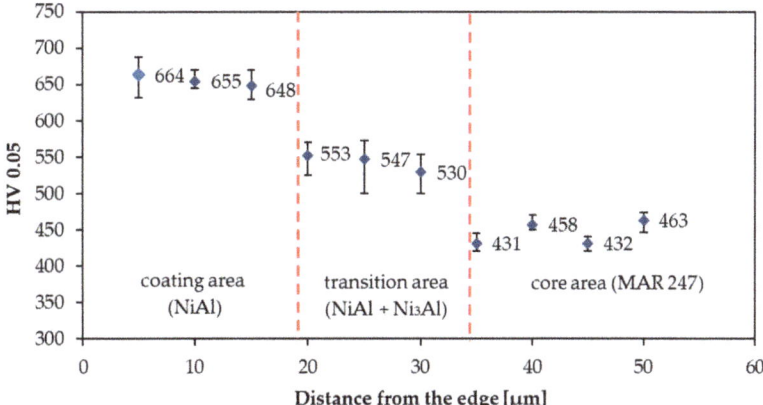

Figure 3. Hardness distribution in the cross-section of coating obtained.

3.2. Effect of Coating on Tensile, Fatigue and Creep Performance of MAR 247 Nickel Based Superalloy

Thermal barrier coatings are playing an increasingly significant role in the advanced gas turbine engines used in aero applications. Current research is focused on development of a new materials for both coatings and turbine blades. The basic properties of these materials are mainly characterized within their microscale using scanning electron microscopy or X-ray diffraction. The mechanical properties are usually evaluated by nanoindentation or microhardness tests. It should be emphasized, however, that the turbine blades are mainly subjected to mechanical loadings at elevated temperature [37]. Unfortunately, not sufficient number of publications was found in this area. Thus, the main aim of work presented here was to perform a wide range of mechanical tests including standard, monotonic tension, fatigue and creep tests to characterized the mechanical response of coated MAR 247 nickel based superalloy at room and elevated temperature.

The uniaxial tensile tests were carried out at room temperature and 1000 °C on coated and uncoated MAR 247 nickel based superalloy. The slight difference of mechanical response was observed for the material tested at room temperature. It has been found, however, that aluminide layer could enhance the high temperature performance of nickel alloy. A considerable improvements of mechanical properties were achieved during testing at 1000 °C (Table 3). The results obtained were similar to those presented in [6] for MAR 247 nickel based superalloy. Moreover, a significant improvement of the yield strength at elevated temperature by approximately of 20% in comparison to Rene 80 [7] and IN738 [8] were found after testing. It should be mentioned, that the coated MAR 247 nickel based superalloy exhibited a comparable mechanical response of such materials as CMSX or Rene 500 [38]. One can indicate, that the aluminide thermal barrier coating used in the conventional MAR 247 nickel based superalloy enhanced the tensile properties of nickel superalloy at high temperature. Moreover, these results suggest, that material tested possessed the equivalent properties to those of the high-strength single crystal superalloys.

Table 3. Selected mechanical properties for coated and uncoated MAR 247 nickel based superalloy after uniaxial tensile test performed at room temperature and 1000 °C.

Material	Room Temperature				1000 °C			
	R_m (MPa)	R_e (MPa)	E (GPa)	A (%)	R_m (MPa)	R_e (MPa)	E (GPa)	A (%)
Uncoated MAR 247 nickel based superalloy	980	800	185	6	380	330	85	5
Coated MAR 247 nickel based superalloy	1100	830	197	6	460	415	98	8

R_m is tensile strength, R_e is yield point, E is Young's modulus, A is elongation.

The effect of coating on mechanical properties of MAR247 nickel based superalloy was investigated in either standard or high temperature fatigue tests. The results shown that coating slightly decrease the elongation of the MAR 247 nickel based superalloy specimens at room temperature while the stress response remained at the similar level (Figure 4a). A significant improvement of mechanical properties was observed during high temperature fatigue testing where the strength of coated specimen increased almost twice in comparison to the as-received material (Figure 4b). Sulak et al. [37] reported that cyclic deformation of nickel-based superalloys can be particularly determined by the interaction analysis between dislocations and γ strengthening phase, where precipitates played a role as an effective barriers against the dislocations movement. Moreover, high temperature performance resulted on phase changes, precipitate coarsening and oxidation [38,39]. All these features led to essential softening of the nickel superalloys.

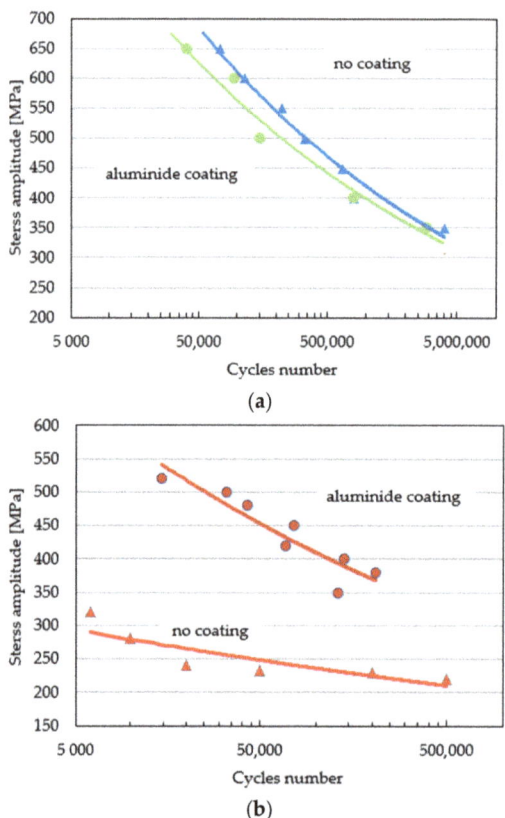

Figure 4. S-N curves of coated and uncoated MAR247 nickel based superalloy determined at: (a) room temperature; (b) 900 °C.

It is found that aluminides coating of 20 μm thickness obtained by CVD process slightly reduced the room temperature performance (Figure 4a). Completely opposite effect was achieved for the same material tested at 900 °C. In this case almost 100% variation of stress amplitude can be easily observed in Figure 4b. Similar strength improvement may be observed in comparison to the conventional, uncoated nickel alloys such as IN 792-5A and IN 713LC during fatigue testing at 900 °C [40]. One can conclude that an application of the aluminide layer can improve significantly the nickel superalloys behavior, and additionally, it can prevent the raw material against such processes as oxidation, hot corrosion, or wear, thus providing a longer life-time [41]. It should be emphasized, that coating obtained in the authors previous study [34] was characterized by the very good adherence, wear and thermal resistance confirmed by experimental studies. Its application significantly improved the hardness and wear in comparison to the as-received MAR 247 nickel based superalloy substrate. The excellent durability and tightness of the protective scale resulted in no scale spallation. It could be concluded, that such aluminide layer considerably enhance the high temperature fatigue performance of MAR 247 nickel based superalloy, and thus, could be successfully applied for nickel alloys. Similar improvement of functional properties of Inconel 713 LC Ni-based superalloy protected with aluminide layer deposited on it during the CVD process was reported in the available papers e.g., [28].

After mechanical testing, the microstructural observations were carried out. They enabled identifications of defects induced due to the loading history defined. Looking at the fracture surfaces, one can indicate a greater or lesser cluster of cracks (Figure 5a,b). Their formation was initiated in the last stage of fatigue, when the strain concentration zone was already formed. This confirms the assumption that cracks in the aluminide layer are preceded by cracks in the entire specimen volume. Such behavior indicates a lower fatigue strength of the layer in comparison to the nickel alloy core.

Figure 5. Microscopic views of the side surfaces of specimens after fatigue tests with visible cracking of the layer in the direction perpendicular to that of deformation: (**a**) edge of the specimen; (**b**) center of the specimen.

The scanning microscope observations of fracture areas of specimens subjected to cyclic loading at room temperature showed that the crack propagation in the layer was not transferred into the nickel matrix. The view of fracture area confirm the permanence of the interfacial adhesion as no cracks between the layer and nickel substrate were observed (Figure 6a). Such behavior can be a sign of good efficiency of CVD process applied. It should be emphasized additionally, that a satisfied layer coherence was attained (Figure 6b). Despite the multiple cracks, the layer remained adhered to the nickel based substrate. Numerous cracks in the aluminide layer were observed near the decohered area (Figure 6c). However, cracks in the layer itself do not limit the possibility of cyclic load carrying. Despite the multiple cracks found on the edge of the sample (Figure 6c), it broke within the gauge. Such behavior may suggest, that the aluminide layer is not responsible for crack initiation. Moreover, it protects the MAR 247 nickel based superalloy against the high temperature exposure, and as a consequence, successfully extends its service life. One can indicate that good coherence between layer and substrate could further improve the mechanical response of matrix material [42].

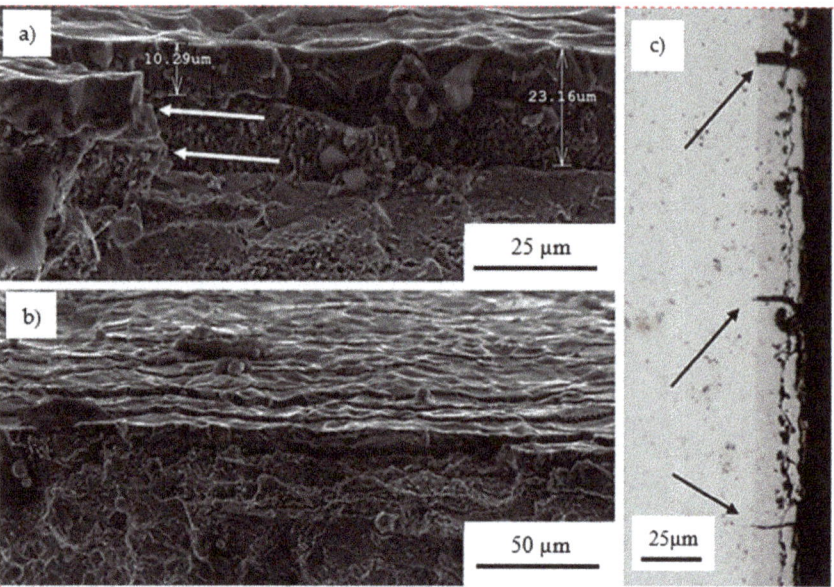

Figure 6. Fracture surface observations: (**a**) propagation of crack through coating without visible cracks in core material; (**b**) morphology of coating after fatigue at room temperature; (**c**) multiple layer cracks.

In the next step of experimental program, the coated and uncoated MAR247 nickel based superalloy was subsequently subjected to creep at 600 °C in the air atmosphere (Figure 7a). It is shown that aluminized layer improved the creep performance of the alloy investigated. The strength enhancement of more than 50 MPa was observed when the high temperature exposure time extended 500 h. The stress response of coated MAR 247 nickel based superalloy tend to decrease steadily with temperature increase while for uncoated material it decreases significantly. It was anticipated that the long time exposure (>1000 h) at high temperature and constant load could even increase the difference in stress response for coated and uncoated material (Figure 7b). Similar work was reported by Yuan et al. [43] where the IN792 in the as-received state and with NiAl protective coating was subjected to creep at temperature of 850 °C and 950 °C. It was concluded, that the crack initiation and propagation mechanisms were temperature dependent. For the polycrystalline superalloy tested at both temperatures, the grain-boundary separation was the main failure mode. In the NiAl coated material, the formation of γ′ along grain boundary in the layer zone at 850 °C could induce active through-coating cracks, which may further penetrate into the substrate along the nearest grain boundary beneath the coating. However, a formation of γ′ in the layer was restricted to 950 °C.

Thermal stability of the coated MAR 247 nickel based superalloy was evaluated through the in-situ Young's modulus measurements in temperature range from 22 to 900 °C during annealing and cooling at a rate of 3 °C/min to 900 °C. RFDA method allowed to monitor an evolution of the Young's modulus with the heating temperature as well as during cooling to room temperature (Figure 8). It was found that such non-destructive method could precisely determine the Young's modulus value in a wide range of temperature. The results were comparable to those obtained from uniaxial tensile tests.

It was concluded that RFDA method could be successfully applied for the non-destructive Young's modulus measurements of coated MAR 247 nickel based superalloy.

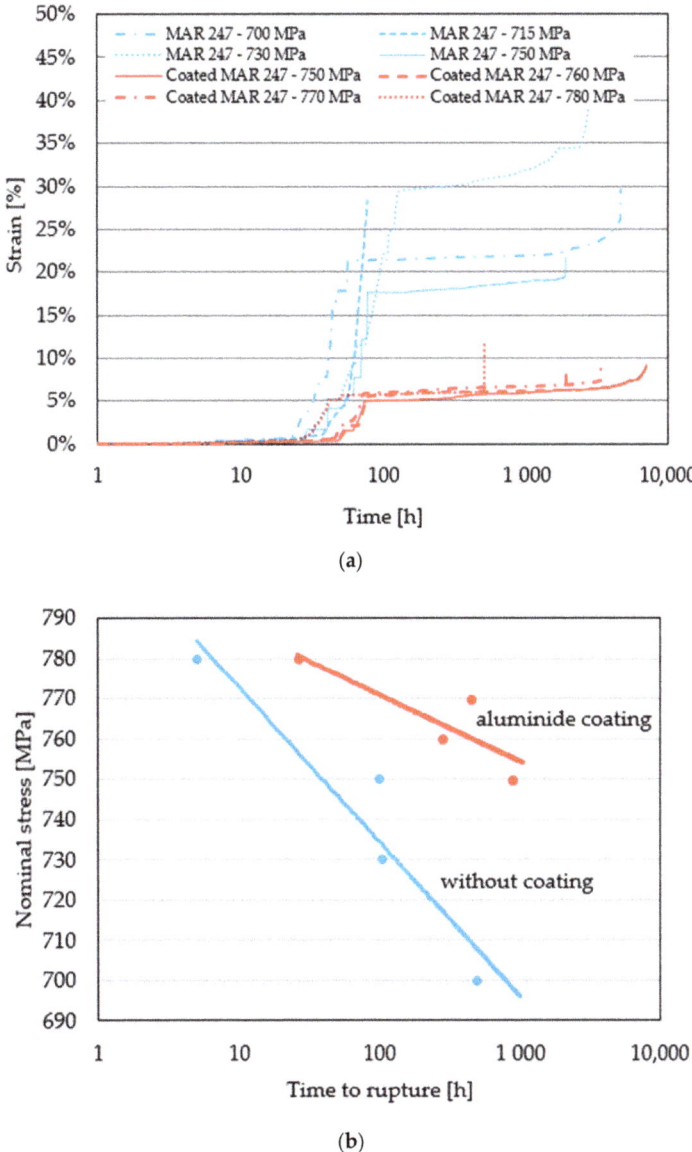

Figure 7. Creep characteristics of coated and uncoated MAR 247 nickel based superalloy (**a**); effect of stress level on time to rupture of coated and uncoated MAR247 nickel based superalloy subjected to creep at 600 °C (**b**).

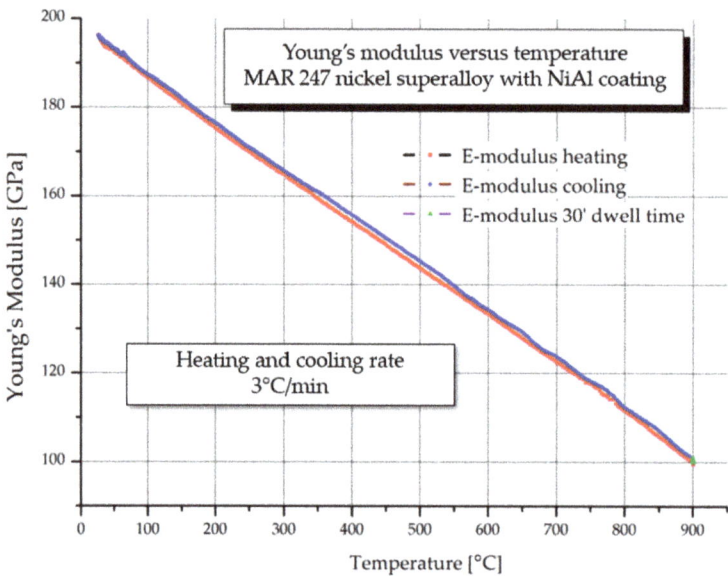

Figure 8. Evolution of the Young's modulus of coated MAR 247 nickel based superalloy at the temperature of 22–900 °C, during annealing and cooling at a rate of 3 °C / min to 900 °C.

4. Conclusions

Optimization of CVD parameters for MAR247 nickel based superalloy with respect to the thermal process applied in the protective hydrogen atmosphere enabled to get an effective, non-defected thermal barrier coating with the uniform thickness. The coating was characterized by a very good adherence and thermal resistance. CVD technology with parameters carefully determined could be successfully used to enhance the mechanical properties of MAR 247 nickel based superalloy. An essential improvement of the strength response during cyclic loading (about 200 MPa) of the coated nickel superalloy in comparison to that of the as-received material achieved was identified at 900 °C. Such feature was not observed at room temperature, where the stress response was almost the same. One can conclude, that the aluminized layer may improve the creep performance of the nickel superalloy significantly. It is evidenced that application of such layer may effectively protect the raw material against such processes as oxidation, hot corrosion or wear, and thus, extend its service life.

Author Contributions: Conceptualization, M.K. and D.K.; methodology, D.K., M.K, X.Y., W.R. and C.S.; formal analysis, M.K., C.S. and Z.L.K.; writing—original draft preparation M.K.; writing—review and editing, C.S. and Z.L.K. All authors have read and agreed to the published version of the manuscript.

Funding: The authors gratefully acknowledge the funding by The National Centre for Research and Development, Poland, under Program for Applied Research, grant no. 178781.

Institutional Review Board Statement: Not applicable.

Informed Consent Statement: Not applicable.

Data Availability Statement: Data available in a publicly accessible repository.

Acknowledgments: The authors are grateful to Mirosław Wyszkowski and Andrzej Chojnacki from the Institute of Fundamental Technological Research of the Polish Academy of Sciences for their great support of the experimental work.

Conflicts of Interest: The authors declare no conflict of interest.

References

1. Agarwal, D.C.; Brill, U. High-temperature-strength nickel alloy. *Adv. Mater. Process.* **2000**, *158*, 31–34.
2. Wanhill, R.J.H. Fatigue of air supply manifold support rod in military jet engines. *JFAP* **2004**, *4*, 53–61. [CrossRef]
3. Goward, G.W. Progress in coatings for gas turbine airfoils. *Surf. Coat. Technol.* **1998**, *1*, 73–79. [CrossRef]
4. Reed, C.R. *The Superalloys-Fundamentals and Applications*; C.U.P.: Cambridge, UK, 2006; ISBN 0511-24546-7.
5. Barbosa, C.; Nascimento, J.L.; Caminha, I.M.V.; Abud, I.C. Microstructural aspects of the failure analysis of nickel base superalloys components. *Eng. Fail. Anal.* **2005**, *12*, 348–361. [CrossRef]
6. Kaufman, M. Properties of cast Mar-M-247 for turbine blisk applications. *Superalloys* **1984**, 43–52. [CrossRef]
7. Fritz, L.J.; Koster, W.P. *Tensile and Creep Rupture Properties of (1) Uncoated and (2) Coated Engineering Alloys at Elevated Temperatures*; NASA Technical Report; NAS CR-135138; NASA: Washington, DC, USA, 1977.
8. NCO. *Alloy IN-738 Technical Data*; The International Nickel Company, Inc.: New York, NY, USA; Available online: https://www.nickelinstitute.org/media/1709/in_738alloy_preliminarydata_497_.pdf (accessed on 1 December 2020).
9. Wee, S.; Do, J.; Kim, K.; Lee, C.; Seok, C.; Choi, B.-G.; Choi, Y.; Kim, W. Review on mechanical thermal properties of superalloys and thermal barrier coating used in gas turbines. *Appl. Sci.* **2020**, *10*, 5476. [CrossRef]
10. Kawahara, Y. Application of high temperature corrosion-resistant materials and coatings under severe corrosive environment in waste-to-energy boilers. *J. Therm. Spray Tech.* **2007**, *16*, 202–213. [CrossRef]
11. Sadeghi, E.; Markocsan, N.; Joshi, S. Advances in corrosion-resistant thermal spray coatings for renewable energy power plants: Part II—Effect of environment and outlook. *J. Therm. Spray Tech.* **2019**, *28*, 1789–1850. [CrossRef]
12. Jozwik, P.; Polkowski, W.; Bojar, Z. Applications of Ni3Al based intermetallic alloys—Current stage and potential perceptivities. *Materials* **2015**, *8*, 2537–2568. [CrossRef]
13. Cinca, N.; Lima, C.R.C.; Guilemany, J.M. An overview of intermetallics research and application: Status of thermal spray coatings. *JMRTAL* **2013**, *2*, 75–86. [CrossRef]
14. Bochenek, K.; Basista, M. Advances in processing of NiAl intermetallic alloys and composites for high temperature aerospace applications. *Prog. Aerosp. Sci.* **2015**, *79*, 136–146. [CrossRef]
15. Curry, N.; Markocsan, N.; Li, X.H.; Tricoire, A.; Dorfman, M. Next generation thermal barrier coatings for the gas turbine industry. *J. Therm. Spray Tech.* **2011**, *20*, 108–115. [CrossRef]
16. Zagula-Yavorska, M.; Sieniawski, J.; Filip, R.; Drajewicz, M. The effect of the aluminide coating on the thermal properties and oxidation resistance of Inconel 625 Ni-base superalloy. *Solid State Phenom.* **2015**, *227*, 313–316. [CrossRef]
17. Stekovic, S. Low cycle fatigue and fracture of a coated superalloy CMSX-4. In *Fracture of Nano and Engineering Materials and Structures*; Gdoutos, E.E., Ed.; Springer: Dordrecht, The Netherland, 2006. [CrossRef]
18. Okazaki, M. High-temperature strength of Ni-base superalloy coatings. *Sci. Technol. Adv. Mat.* **2001**, *2*, 357–366. [CrossRef]
19. Rodriguez, P.; Mannan, S.L. High temperature low cycle fatigue. *Sadhana* **1995**, *20*, 123–164. [CrossRef]
20. Castillo, R.; Willett, K.P. The effect of protective coatings on the high temperature properties of a gamma prime-strengthened Ni-base superalloy. *Metall. Mater. Trans. A* **1984**, *15*, 229–236. [CrossRef]
21. Veys, J.M.; Mevrel, R. Influence of protective coatings on the mechanical properties of CMSX-2 and Cotac 784. *Mater. Sci. Eng. A* **1987**, *88*, 253–260. [CrossRef]
22. Kalivodova, J.; Baxter, D.; Schutze, M.; Rohr, V. Corrosion behaviour of boiler steels, coatings and welds in flue gas environments. *Mater. Corros.* **2008**, *59*, 367–373. [CrossRef]
23. Kochmańska, A.; Garbiak, M. High-temperature diffusion barrier for Ni–Cr Cast Steel. *Defect Diffus. Forum.* **2011**, 595–600. [CrossRef]
24. Zhan, Z.; He, Y.; Li, L.; Liu, H.; Dai, Y. Low-temperature formation and oxidation resistance of ultrafine aluminide coatings on Ni-base superalloy. *Surf. Coat. Technol.* **2009**, *203*, 2337–2342. [CrossRef]
25. Xu, Z.H.; Dai, J.W.; Niu, J.; He, L.M.; Mu, R.D.; Wang, Z.K. Isothermal oxidation and hot corrosion behaviors of diffusion aluminide coatings deposited by chemical vapor deposition. *J. Alloys Compd.* **2015**, *637*, 343–349. [CrossRef]
26. Wang, K.L.; Chen, F.S.; Leu, G.S. The aluminizing and Al–Si codeposition on AISI HP alloy and the evaluation of their carburizing resistance. *Mater. Sci. Eng. A* **2003**, *357*, 27–38. [CrossRef]
27. Wang, Y.; Chen, W. Microstructures, properties and high-temperature car-burization resistances of HVOF thermal sprayed NiAl intermetallic-based alloy coatings. *Surf. Coat. Technol.* **2004**, *183*, 18–28. [CrossRef]
28. Yavorska, M.; Sieniawski, J.; Zielińska, M. Functional properties of aluminide layer deposited on inconel 713 lc ni-based superalloy in the CVD process. *Arch. Metall. Mater.* **2011**, *56*, 187–192. [CrossRef]
29. Bojar, Z.; Jóźwik, P.; Bystrzycki, J. Tensile properties and fracture behavior of nanocrystalline Ni3Al intermetallic foil. *Scr. Mat.* **2006**, *55*, 399–402. [CrossRef]
30. Polkowski, W.; Jóźwik, P.; Karczewski, K.; Bojar, Z. Evolution of crystallographic texture and strain in a fine-grained Ni3Al (Zr, B) intermetallic alloy during cold rolling. *Arch. Civ. Mech. Eng.* **2014**, *14*, 550–560. [CrossRef]
31. Adamiak, S.; Bochnowski, W.; Dziedzic, A.; Filip, R.; Szeregij, E. Structure and properties of the aluminide coatings on the Inconel 625 superalloy. *High. Temp. Mat. Pr. Isr.* **2016**, *35*, 103–112. [CrossRef]
32. Ning, B.; Stevenson, M.E.; Weaver, M.L.; Bradt, R.C. Apparent indentation size effect in a CVD aluminide coated Ni-base superalloy. *Surf. Coat. Technol.* **2003**, *163*, 112–117. [CrossRef]
33. Zagula-Yavorska, M.; Kocurek, P.; Pytel, M. Oxidation resistance of turbine blades made of ŻS6K superalloy after aluminizing by Low-Activity CVD and VPA Methods. *J. Mater. Eng. Perform.* **2016**, *25*, 1964–1973. [CrossRef]

34. Kukla, D.; Kopec, M.; Kowalewski, Z.L.; Politis, D.J.; Jozwiak, S.; Senderowski, C. Thermal barrier stability and wear behavior of CVD deposited aluminide coatings for MAR 247 nickel superalloy. *Materials* **2020**, *12*, 3863. [CrossRef] [PubMed]
35. Maxwell, A.S.; Owen-Jones, S.; Jennett, N.M. Measurement of Young's modulus and Poisson's ratio of thin coatings using impact excitation and depth-sensing indentation. *Rev. Sci. Instrum.* **2004**, *75*, 970–975. [CrossRef]
36. ASTM E1876-99. *Standard Test Method for Dynamic Young's Modulus, Shear Modulus, and Poisson's Ratio by Impulse Excitation of Vibration*; ASTM International: West Conshohocken, PA, USA, 2001.
37. Sulak, I.; Obrtlik, K.; Celko, L. High-temperature low-cycle fatigue behaviour of HIP treated and untreated superalloy MAR-M247. *Kovove Mater.* **2016**, *54*, 471–481. [CrossRef]
38. Corrigan, J.; Launsbach, M.G.; Mihalisin, J.R. Nickel Base Superalloy and Single Crystal Castings. U.S. Patent 8,241,560, 14 August 2012.
39. Antolovich, S.D. Microstructural aspects of fatigue in Ni-base superalloys. *Phil. Trans. R. Soc. A* **2015**, *373*, 20140128. [CrossRef] [PubMed]
40. Šmíd, M.; Kunz, L.; Hutař, P.; Hrbáček, K. High cycle fatigue of nickel-based superalloy MAR-M 247 at high temperatures. *Proc. Eng.* **2014**, *74*, 329–332. [CrossRef]
41. Rahmani, K.; Nategh, S. Mechanical properties of uncoated and aluminide-coated René 80. *Metall. Mater. Trans. A Phys. Metall. Mater. Sci.* **2010**, *41*, 125–137. [CrossRef]
42. Samal, S.; Tyc, O.; Heller, L.; Šittner, P.; Malik, M.; Poddar, P.; Catauro, M.; Blanco, I. Study of interfacial adhesion between nickel-titanium shape memory alloy and a polymer matrix by laser surface pattern. *Appl. Sci.* **2020**, *10*, 2172. [CrossRef]
43. Yuan, K.; Peng, R.; Li, X.-H.; Johansson, L.; Johansson, S.; Wang, Y. Creep fracture mechanism of polycrystalline Ni-based superalloy with diffusion coatings. In Proceedings of the 13th International Conference on Fracture ICF13, Beijing, China, 16–21 June 2013. Available online: http://urn.kb.se/resolve?urn=urn:nbn:se:liu:diva-8949 (accessed on 1 December 2020).

Article

Study on Corrosion Resistance and Wear Resistance of Zn–Al–Mg/ZnO Composite Coating Prepared by Cold Spraying

Xinqiang Lu [1], Shouren Wang [1,*], Tianying Xiong [2], Daosheng Wen [1], Gaoqi Wang [1] and Hao Du [2]

1. School of Mechanical Engineering, University of Jinan, Jinan 250022, China
2. Institute of Metal Research, Chinese Academy of Sciences, Shenyang 110016, China
* Correspondence: me_wangsr@ujn.edu.cn

Received: 13 July 2019; Accepted: 7 August 2019; Published: 9 August 2019

Abstract: Two composite coatings, $Zn_{65}Al_{15}Mg_5ZnO_{15}$ and $Zn_{45}Al_{35}Mg_5ZnO_{15}$, were prepared by the cold spray technique and were found to be compact, with no pits or cracks, based on scanning electron microscope (SEM) and energy-dispersive X-ray spectroscopy (EDS) investigations. The results of the neutral salt spray (NSS) and electrochemical tests showed that the two composite coatings possess a suitable corrosion performance. However, the $Zn_{45}Al_{35}Mg_5ZnO_{15}$ composite coatings were more corrosion resistant and allowed a better long-term stability. In addition, they were found to exhibit the best wear resistance and photocatalytic degradation efficiency.

Keywords: cold spray; scanning electron microscope; electrochemical workstation; neutral salt spray test; photocatalysis; friction and wear; composite coatings

1. Introduction

As a result of the larger development and utilization of marine resources, metallic materials representing the main building elements of the engineering equipment operating in the marine environment are strongly affected by its corrosive nature, significantly shortening their lifetime. Therefore, increased interest has been given to the development of efficient protective coatings suitable to be applied on the metallic parts of offshore equipment [1–4].

Currently, the commonly used surface treatment procedures for equipment working in the marine environment include thermal spraying, hot dip coating, and organic coating [5–8]. However, several drawbacks have been noticed, including an uncontrolled oxidation and increase of the pores within the coating during thermal spraying, as well as a faster degradation of the organic coatings due to their low UV resistance, releasing toxic compounds.

1.1. Cold Spray Technology Features

Cold gas dynamic spray, also known as cold gas spray, is a new surface treatment technology which has been rapidly developed in recent years. It uses compressed gas (air, helium, nitrogen, etc.) as an accelerated gas stream to drive powder particles (particle size 1–50 µm) to collide with the substrate at low temperature (room temperature to 600 °C), supersonic speed (300–1200 m/s), and complete solid state so that the particles undergo strong plastic deformation and deposit to form a coating [9–14].

Compared to the traditionally used treatment processes for offshore equipment protection, cold spray technique has the following advantages: (1) the coating prepared by cold spraying technology is in a low-temperature environment, the deposition rate of the coating is high, and it can be applied to various surface treatments of substrates, (2) it has little effect on the heat of the raw powder particles and the substrate, (3) the prepared coating has a compact structure and low porosity, and (4) the coating prepared by cold spraying has a strong ability to protect the matrix. On the one hand, the coatings have the advantages of original coatings, which can physically isolate seawater from the matrix. On the other hand, even if the coatings are scratched, they can effectively protect the matrix of marine equipment from sacrificial anode electrochemical protection.

1.2. Cold Spray Coating on the Surface of Marine Equipment

Studies show that Zn coating [15,16] is the best electrochemical protection for matrix steel, but its stability, long-term effectiveness, and erosion resistance in a high chlorine environment are not as good as that of Al coating [17–19]. Therefore, Zn–Al coatings [20–24], Zn–Mg coatings [25], Zn–Ni coatings [26,27], and other composite coatings have emerged as the times require. They all have good corrosion resistance and compact structures. The addition of Mg leads to changes in the coating structure and the corrosion product film, and the Zn–Mg coating has a better corrosion resistance than the pure Zn coating. Among them, Zn–6Al–3Mg is the most corrosion resistant of the zinc-based alloys studied, and its corrosion resistance is 18 times that of Zn coating [28–30]. Photocatalytic technology has been rapidly developed due to its wide and complete reaction conditions, wide decomposition range, and low energy consumption cost. Adding ZnO with good photocatalytic performance to a Zn-based alloy can play the role of double prevention. Seawater corrosion is also resistant to marine microbial fouling, and it is the best protection system for marine corrosion and biofouling protection.

For further study, two new types of Zn–Al–Mg/ZnO composite coatings, known as $Zn_{65}Al_{15}Mg_5ZnO_{15}$ and $Zn_{45}Al_{35}Mg_5ZnO_{15}$, were designed on the basis of Zn–Al coating.

2. Experimental Methods

2.1. Preparation of Coating

Spherical powder particles of Zn, Mg, Al, and ZnO having sizes in the range of 10–30 μm were used to prepare the composite coatings. In order to synthesize the $Zn_{65}Al_{15}Mg_5ZnO_{15}$ and $Zn_{45}Al_{35}Mg_5ZnO_{15}$ coatings, the corresponding mass percentages of each component were mixed. Then they were mixed in a ball mill for 1.5 h to obtain uniform coating spraying materials. The matrix was made of 20 mm × 100 mm × 2 mm Q235 steel (Q235 is a carbon structural steel, also called level of steel. Q represents the yield limit of this material, and the latter 235 refers to the yield value of this material, which is around 235 MPa.). Before spraying, the surface of the Q235 steel was strengthened by sand blasting and descaling.

The composite coatings were prepared using a cold spraying technique (DyMET 423). The core of the cold spraying equipment is the spray gun mechanism, which is equipped with a high-pressure gas pipeline, heater, powder feeding pipeline, and de Laval nozzle (diameter 2 mm). The whole spray gun mechanism is installed on a six-degrees-of-freedom manipulator arm. The compressed power gas is nitrogen at the pressure of 1.8 MPa, and its working temperature is 400 °C. The distance between the spray gun and the spraying base is 20 mm, and the spraying speed of the powder particles is 800 m/s. Figure 1 is a schematic diagram of the working principle of the cold spray coating preparation.

The coatings were cut into 10 mm × 10 mm samples by laser cutting technology. The sample surface was polished repeatedly, then the impurities were removed by ultrasonic cleaning in absolute ethanol. Finally, the treated samples were stored in a vacuum bag for further tests.

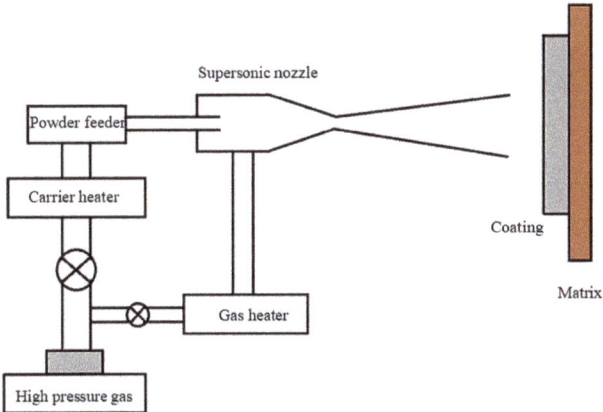

Figure 1. The schematic diagram of the working principle of the cold spray coating preparation.

2.2. Coating Characteristics

The microstructures of the composite coatings of $Zn_{65}Al_{15}Mg_5ZnO_{15}$ and $Zn_{45}Al_{35}Mg_5ZnO_{15}$ were observed and characterized. On the one hand, scanning electron microscopy (SEM, JSM-7610F, JEOL, Tokyo, Japan) provided information on the coating morphology and further summarized the microstructure characteristics of the coating. On the other hand, the elemental composition of the coating was characterized by an energy-dispersive X-ray spectroscopy (EDS) function of a scanning electron microscope (SEM). The parameters set by the scanning electron microscope to observe the microstructure of the coating areas was as follows: continuous scanning mode, scanning speed of 2°/min, scanning range of 10° to 90°, acceleration voltage of 20 kV, resolution of 1 nm, and so on.

2.3. Corrosion Behavior of the Composite Coatings

The neutral salt spray test (NSS, LRHS-108-RY, Shanghai Forestry Instrument Co., Ltd., Shanghai, China) was used to assess the corrosion behavior in the presence of a chloride aggressive environment. In order to assess the corrosion performance of the two composite coatings involving NSS, the selected specimens (area of 10 mm × 10 mm) were divided into four groups, each with four samples. The grouped samples were placed on the salt spray test box, waiting for the start of the neutral salt spray test. The parameters of the neutral salt spray test were as follows: 1000 mL of 3.5% ± 1.5% NaCl solution was placed into the spray chamber, the pH of the NaCl solution was between 6.5 and 7.2, the working temperature of the salt spray test chamber was stable at 35 ± 2 °C, the neutral salt spray test work process was static at room temperature, the NSS test for the investigated composite coatings had a total duration of 480 h, and intermediary examinations were performed after 120, 240, and 360 h of conditioning.

In addition, electrochemical accelerated tests were also performed using a 3.5% ± 1.5% NaCl solution, involving an electrochemical workstation (CHI604E, Shanghai Chenhua Instruments Co., Ltd., Shanghai, China). A three-electrode electrochemical cell was used, where Ag–AgCl (in saturated KCl) was the reference electrode, a Pt plate (10 mm × 10 mm) was the auxiliary electrode, and the composite coating sample was the working electrode. Prior to the electrochemical accelerated corrosion test, the non-coated surface of the coated sample was sealed with Kraft silicone rubber to prevent other surfaces affecting the coating test. The samples of the two composite coatings were divided into five labeled groups, and the labeled samples were separately placed in a 3.5% ± 1.5% NaCl solution where the solution was intermittently magnetically stirred every 12 h. The five groups of composite coating samples were soaked for 1, 120, 240, 360, and 480 h, respectively. The immersed composite coatings were placed in the electrochemical cell to measure their stable open circuit potential and corresponding polarization curves. The open circuit potential and polarization curves of each set of samples were measured multiple times to obtain good open circuit potential and plan curves. Then a Tafel curve test was performed at a scanning rate of 0.5 mV/s. A Butler–Volmer analysis involving Thales XT5.1.4 software was used to determine the corrosion current values (denoted *Icorr*) for each investigated working electrode. In addition, the polarization curves of Q235 matrix steel were compared with those of $Zn_{65}Al_{15}Mg_5ZnO_{15}$ and $Zn_{45}Al_{35}Mg_5ZnO_{15}$ composite coatings.

2.4. Photocatalytic Characteristics of the Composite Coatings

The photocatalytic degradation characteristics of $Zn_{65}Al_{15}Mg_5ZnO_{15}$ composite coating, $Zn_{45}Al_{35}Mg_5ZnO_{15}$ composite coating, and matrix steel Q235 on methyl blue were studied by UV spectrophotometer (UV-5100, Shanghai Yuanfang Instrument Co., Ltd., Shanghai, China). Samples were placed in a beaker containing methyl blue solution with a concentration of 100 mL and 20 mg/L. Ultraviolet light was directly irradiated on the coating surface. In the experiment, a magnetic stirrer (CL-200, Gongyi Yuhua Instrument Co., Ltd, Zhengzhou, China) was used to stir the solution continuously to prevent degradation products remaining on the coating surface from affecting the sustainability of degradation. Every 5 or 10 min, the pipette was used to put 3 mL solution into the colorimetric tube. The absorbance of the solution was measured by ultraviolet spectrophotometer, and the concentration of methyl blue in each time period was obtained. When degrading, the ultraviolet light should be irradiated on the surface of the coating as much as possible to ensure that the degradation efficiency of the coating is not affected. A comparative analysis of the photodegradation of methyl blue under UV irradiation on the investigated composite coatings and on Q235 steel was performed.

The degradation rate was calculated using Equation (1) below. The degradation rate of methyl blue under different materials was calculated by photocatalytic degradation test data.

$$\eta = (A_0 - A_t)/A_0 \times 100\% \tag{1}$$

where η is the degradation rate, A_0 is the initial absorbance of the maximum absorption peak of the methyl blue solution, and A_t is the absorbance of the maximum absorption peak at the time, t, of the methyl orange solution.

2.5. Composite Coating Friction and Wear Test

The friction and wear characteristics of the investigated composite coating were tested by a friction and wear tester (MFT-50, Rtec Instruments, San Jose, CA, USA). The friction mode adopts spherical reciprocating wet friction, wherein the friction pair (the workpiece to be ground) is a GCr15 bearing steel ball (MFT-50, Rtec Instruments, San Jose, CA, USA) having a diameter of 6.35 mm. The relevant parameters were set according to the conditions of the material of the composite coating. The reciprocating friction frequency was 4 Hz, the grinding stroke was 4.50 mm, the friction and wear positive pressure was 20 N, and the friction and wear test time was 90 min. The entire friction and wear test was carried out in a 3.5% NaCl solution environment with a relative humidity of 45% ± 5%. The 3D morphology characterizing the frictional wear surface was observed using a white light interferometer (MFT-5000, Rtec Instruments, San Jose, CA, USA). The samples were weighed separately before and after the friction and wear test, and the wear and wear rates of the samples were calculated.

3. Results and Discussion

3.1. Preparation of Composite Coatings

Figure 2 presents the images of the $Zn_{65}Al_{15}Mg_5ZnO_{15}$ and $Zn_{45}Al_{35}Mg_5ZnO_{15}$ composite coated sheets prepared by cold spraying technology. It can be observed that the surfaces of the composite coated plates of $Zn_{65}Al_{15}Mg_5ZnO_{15}$ and $Zn_{45}Al_{35}Mg_5ZnO_{15}$ are very smooth and uniform. However, it can also be seen that the surface of the $Zn_{65}Al_{15}Mg_5ZnO_{15}$ composite coating has obvious pits and defects. After several sprays, the $Zn_{65}Al_{15}Mg_5ZnO_{15}$ composite coating still had obvious pits and defects, whereas the surface of the $Zn_{45}Al_{35}Mg_5ZnO_{15}$ composite coating was very flat and smooth, and the coating surface had no obvious defects. By comparing the macroscopical appearance of the $Zn_{65}Al_{15}Mg_5ZnO_{15}$ and $Zn_{45}Al_{35}Mg_5ZnO_{15}$ composite coatings prepared by cold spraying, it was found that the spraying deposition effect of $Zn_{45}Al_{35}Mg_5ZnO_{15}$ was superior to that of $Zn_{65}Al_{15}Mg_5ZnO_{15}$.

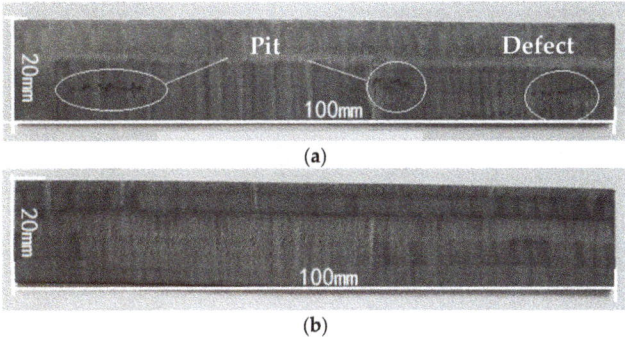

Figure 2. Surface characteristics of the $Zn_{65}Al_{15}Mg_5ZnO_{15}$ (**a**) and $Zn_{45}Al_{35}Mg_5ZnO_{15}$ (**b**) composite coatings taken with an optical camera.

3.2. Composite Coating Morphology

Figure 3 shows the morphology of the surface of typical $Zn_{45}Al_{35}Mg_5ZnO_{15}$ and $Zn_{65}Al_{15}Mg_5ZnO_{15}$ composite coatings. There are three main morphology characteristics of the composite coatings: bright spots and white and black regions. Figure 3a,d is the enlarged 300× micromorphology of the composite coatings, Figure 3b,e is the enlarged 3000× micromorphology of the composite coatings, and Figure 3c,f is the enlarged 300× micromorphology of the cross-section of the composite coatings. Figure 3 shows the micromorphology of the composite coatings. The composite coatings have three main components, which are uniformly distributed in the coatings, and the same appearance does not have a large amount of aggregation. It is further explained that the mechanical mixing of cold sprayed coatings Zn, Al, Mg, and ZnO is very uniform in the configuration process. Combining the micromorphology and micro cross-sectional structure of the two composite coatings, it was found that the three morphology characteristics were uniform strips and closely adhered to each other to form a compact coating. There were no obvious defects, such as pits, in the internal structure of the composite coatings.

Figures 4 and 5 present the energy dispersion point spectra of the three kinds of composite elements in the microstructures of the $Zn_{45}Al_{35}Mg_5ZnO_{15}$ and $Zn_{65}Al_{15}Mg_5ZnO_{15}$ composite coatings. Figures 4a and 5a show the microstructures of the two composite coatings magnified 3000×. Figures 4b–d and 5b–d are the energy-dispersive spectra of Points 1, 2, and 3 corresponding to the three appearances of the composite coating surface, respectively. Energy is the element content of a point represented by a dispersion spectrogram. By comparing the energy dispersion spectra of the two layers in Figures 4 and 5, it was found that they have similarity in structure, morphology, and element content. The elemental composition of Points 1, 2, and 3 was substantially similar, so the components of the two coatings were analyzed together. In Figures 4b and 5b, the Al content in Point 1 is 91.8% and 90.1%, and the Zn content is only 8.2% and 9.9%, so the element of the black region in the overcoat layer could be analyzed. The Zn content in Point 2 in Figures 4c and 5c is 98.8% and 98.6%, but the Al content is only 1.2% and 1.4%, so the white area in the composite coating could be obtained by elemental content analysis. The element is Zn. Figures 4d and 5d contain only Zn and O elements, so the element in the bright spot area is ZnO.

Through the analysis and observation of two kinds of composite coatings from two perspectives (plane and cross-section), shown in Figure 3, it was found that obvious plastic deformation occurs in the preparation of powder coatings and that the overall structure is flat. The structure of the composite coating is compact, and there are no pits and defects in the surface or internal structure.

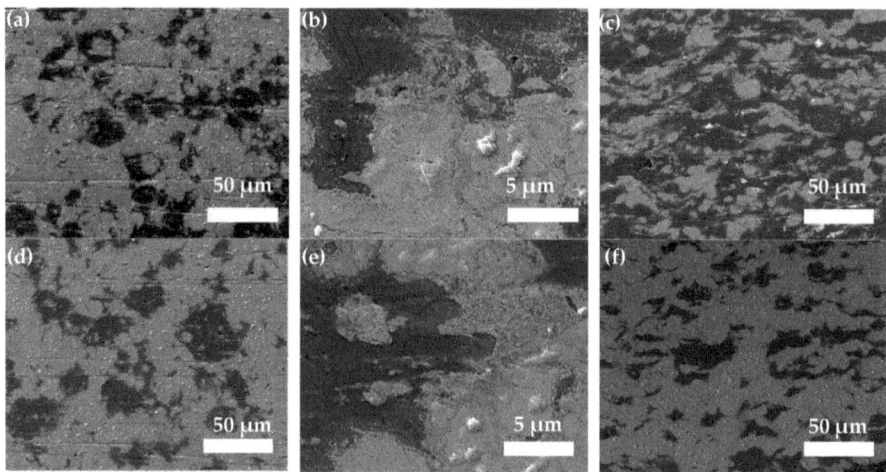

Figure 3. Microstructure of the surface of two composite coatings: (**a**) the surface of the $Zn_{45}Al_{35}Mg_5ZnO_{15}$ composite coating (magnification 300×), (**b**) the surface of the $Zn_{45}Al_{35}Mg_5ZnO_{15}$ composite coating (magnification 3000×), (**c**) $Zn_{45}Al_{35}Mg_5ZnO_{15}$ composite coating cross-section (magnification 300×), (**d**) the surface of the $Zn_{65}Al_{15}Mg_5ZnO_{15}$ composite coating (magnification 300×), (**e**) the surface of the $Zn_{65}Al_{15}Mg_5ZnO_{15}$ composite coating (enlargement 3000×), and (**f**) $Zn_{65}Al_{15}Mg_5ZnO_{15}$ composite coating section (magnification 300×).

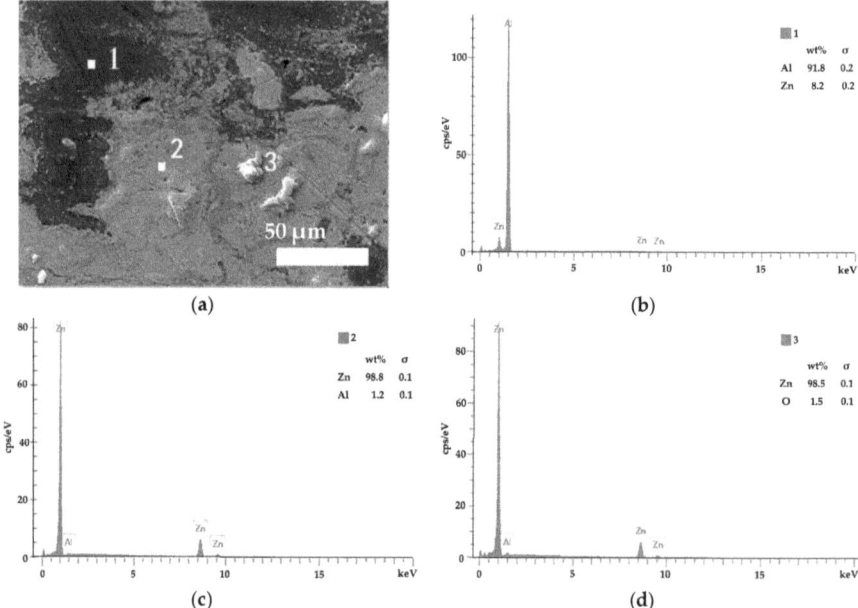

Figure 4. (**a**) Energy dispersion point spectrum (EDSP) of the $Zn_{45}Al_{35}Mg_5ZnO_{15}$ composite coatings amplified at 3000× microstructure for the element analysis of three specimens: (**b**) Point 1, (**c**) Point 2, and (**d**) Point 3.

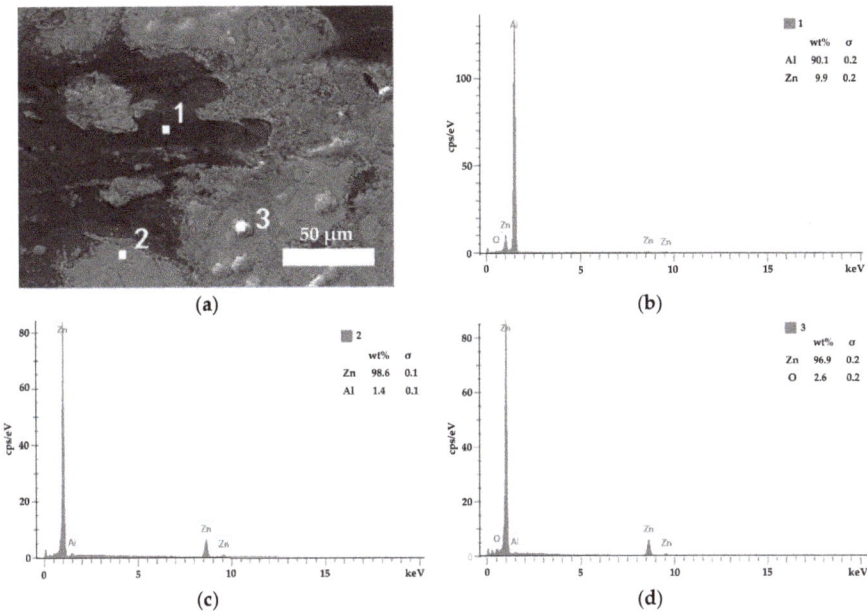

Figure 5. Energy dispersion point spectrum (EDSP) of the $Zn_{65}Al_{15}Mg_5ZnO_{15}$ composite coatings amplified at 3000× microstructure for the element analysis of three specimens: (**b**) Point 1, (**c**) Point 2, and (**d**) Point 3.

3.3. Corrosion Resistance Test

3.3.1. Electrochemical Accelerated Corrosion Test

The dependence of the open circuit potential against time provides information on corrosion evolution. The displacement of the corrosion potential toward more electronegative values suggests an increase in the corrosion rate. Figure 6 presents the evolution of the open circuit potential against the time for the two investigated composite coatings. Figure 6 shows the more negative displacement of the potential of the two composite coatings at the onset of corrosion. When the etching time is 120 h, the corrosion open circuit potential of the coating gradually shifts to the positive direction until it is maximum. After 360 h, the open circuit potentials of the two composite coatings began to shift in the negative direction. During the whole corrosion process, the corrosion process of the $Zn_{65}Al_{15}Mg_5ZnO_{15}$ and $Zn_{45}Al_{35}Mg_5ZnO_{15}$ composite coatings was basically the same, and the open circuit potential was synchronously shifted in the positive direction or negative direction. However, when the corrosion is stable, the open circuit potential of the $Zn_{45}Al_{35}Mg_5ZnO_{15}$ composite coating is always displaced in the positive direction of the $Zn_{65}Al_{15}Mg_5ZnO_{15}$ composite coating.

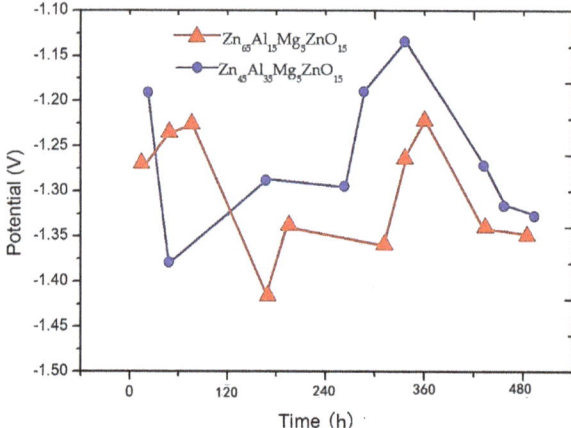

Figure 6. Open circuit potentials of $Zn_{65}Al_{15}Mg_5ZnO_{15}$ and $Zn_{45}Al_{35}Mg_5ZnO_{15}$ at different times.

Figure 7 shows the polarization curves of the two kinds of composite coatings at 1, 24, 120, 240, 360, and 480 h and the polarization curves of matrix steel Q235. Table 1 shows the values of the corrosion potential and of the corrosion current density for the two composite coatings at 1, 120, 360, and 480 h and the polarization potential and corrosion current density of Q235. Figure 7a is a comparison of the polarization curves of the $Zn_{45}Al_{35}Mg_5ZnO_{15}$ composite coating at different times and the planned curve of the matrix steel. It was found that the polarization potential of the $Zn_{45}Al_{35}Mg_5ZnO_{15}$ composite coating at different times (maximum −1.32 V) is always much smaller than that of the base steel Q235. The polarization potential (−0.95 V) reveals that the $Zn_{45}Al_{35}Mg_5ZnO_{15}$ composite coating has a strong anodizing protection of the base steel in a seawater environment. Figure 7b is a comparison of the polarization curves of the $Zn_{65}Al_{15}Mg_5ZnO_{15}$ composite coating at different times and the planned curve of the matrix steel. It was found that the polarization potential of the $Zn_{65}Al_{15}Mg_5ZnO_{15}$ composite coating at different times (maximum −1.22 V) is always much smaller than the base steel Q235. The polarization potential (−0.95 V) shows that the $Zn_{65}Al_{15}Mg_5ZnO_{15}$ composite coating also has strong anodized protective base steel properties in the seawater environment. By comparing the corrosion current densities of the two composite coatings in Table 1, it was found that the corrosion current density of the $Zn_{65}Al_{15}Mg_5ZnO_{15}$ composite coating is higher than that of the $Zn_{45}Al_{35}Mg_5ZnO_{15}$ composite coating under the same corrosion time. In other words, the corrosion rate of the $Zn_{45}Al_{35}Mg_5ZnO_{15}$ composite coating always displaced in the more positive direction as compared to the $Zn_{65}Al_{15}Mg_5ZnO_{15}$ composite coating.

Table 1. Polarization potential and corrosion current density of two composite coatings at 1, 120, 360, and 480 h, and the polarization potential and corrosion current density of Q235.

Samples	Times/h	Ecorr (V)	Icorr (A·cm^{-2})
Q235	1	−0.95 ± 0.02	6.58×10^{-4}
$Zn_{45}Al_{35}Mg_5ZnO_{15}$	1	−1.22 ± 0.02	2.11×10^{-5}
	120	−1.42 ± 0.02	1.08×10^{-5}
	360	−1.27 ± 0.02	1.28×10^{-4}
	480	−1.20 ± 0.02	8.23×10^{-5}
$Zn_{65}Al_{15}Mg_5ZnO_{15}$	1	−1.41 ± 0.02	3.03×10^{-5}
	120	−1.50 ± 0.02	1.63×10^{-5}
	360	−1.32 ± 0.02	3.54×10^{-4}
	480	−1.40 ± 0.02	1.08×10^{-4}

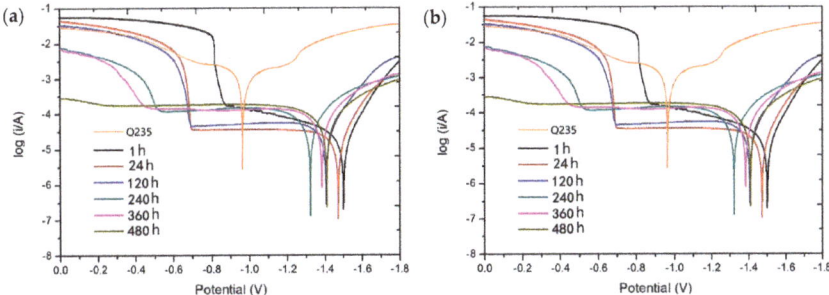

Figure 7. Polarization curves of two composite coatings at 1, 24, 120, 240, 360, and 480 h and polarization curves of matrix steel Q235: (**a**) polarization curves of $Zn_{45}Al_{35}Mg_5ZnO_{15}$ at different times and (**b**) polarization curves of $Zn_{65}Al_{15}Mg_5ZnO_{15}$ at different times.

Based on the data of Table 1 and Figures 6 and 7, it was found that both the $Zn_{65}Al_{15}Mg_5ZnO_{15}$ and $Zn_{45}Al_{35}Mg_5ZnO_{15}$ composite coatings have good properties of anodic oxidation protection of matrix steel. However, the open circuit potential and corrosion current density show that the anodic oxidation rate of the $Zn_{65}Al_{15}Mg_5ZnO_{15}$ composite coating is always higher than that of the $Zn_{45}Al_{35}Mg_5ZnO_{15}$ composite coating. Therefore, considering the time effect of protecting matrix steel, the composite coating of $Zn_{45}Al_{35}Mg_5ZnO_{15}$ has a longer durability than that of $Zn_{65}Al_{15}Mg_5ZnO_{15}$ under the same coating conditions.

3.3.2. Neutral Salt Spray Test of Composite Coating

Figure 8 presents the SEM micrographs of the $Zn_{45}Al_{35}Mg_5ZnO_{15}$ composite coatings after different conditioning periods. Figure 8a shows the micromorphology of the composite coating after 120 h of corrosion. A dense network morphology is formed on the surface of the coating after corrosion. Figure 8b shows the micromorphology of the composite coating after 240 h of corrosion. The surface of the coating is corroded to form a dense flocculent corrosion product. The interior of the corrosion product is a lamellar material. Figure 8c shows the micromorphology of the composite coating after 360 h of corrosion. The flocculent morphology products on the surface of the coating are gradually eroded away, and the second layer of the dense flocculent morphology is formed by corrosion. Figure 8d shows the micromorphology of the composite coating after 480 h of corrosion, and the flocculent structure on the surface of the coating continues to corrode. The results show that the corrosion of the $Zn_{45}Al_{35}Mg_5ZnO_{15}$ composite coating is regularly stratified and that there are no pits and cracks in the corrosion process.

Figure 9 presents the SEM micrographs of the $Zn_{65}Al_{15}Mg_5ZnO_{15}$ composite coatings after different conditioning periods. Figure 9a shows the micromorphology of the composite coating after 120 h of corrosion. The surface of the coating is corroded to form a dense flocculent corrosion product. The interior of the corrosion product is a lamellar material. Figure 9b shows the micromorphology of the composite coatings after 240 h of corrosion. The deeper coatings begin to corrode irregularly, and deep pits appear on the surface of the corrosion products. Figure 9c shows the micromorphology of the composite coating after 360 h of corrosion. The surface of the coating is etched and cracks and pits appear. Figure 9d shows the micromorphology of the composite coating etched for 480 h—the surface of the coating is severely corroded and severe cracks and pits appear. The results show that the corrosion of the $Zn_{65}Al_{15}Mg_5ZnO_{15}$ composite coating is not uniform and that deep pits and cracks are produced after 240 h of corrosion.

Figures 8 and 9 show that the $Zn_{45}Al_{35}Mg_5ZnO_{15}$ composite coating has regular delamination corrosion and that no pits and crack defects occur during the corrosion process. The $Zn_{65}Al_{15}Mg_5ZnO_{15}$ composite coating is not uniformly etched and defects, such as pits and cracks, appear. Therefore, the

$Zn_{45}Al_{35}Mg_5ZnO_{15}$ composite coating has a better corrosion resistance than the $Zn_{65}Al_{15}Mg_5ZnO_{15}$ composite coating under the same corrosion conditions.

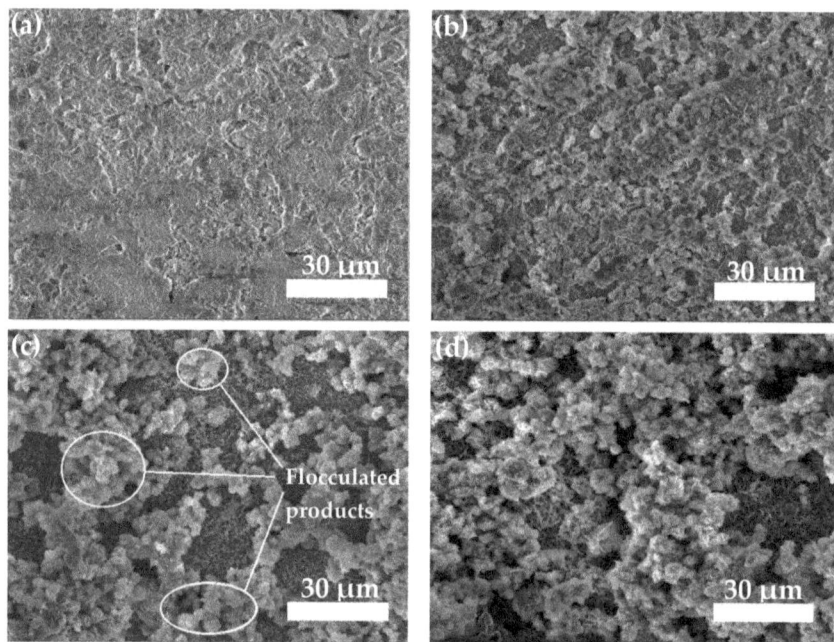

Figure 8. SEM micrographs for the $Zn_{45}Al_{35}Mg_5ZnO_{15}$ coating after different conditioning periods: (**a**) 120 h, (**b**) 240 h, (**c**) 360 h, and (**d**) 480 h.

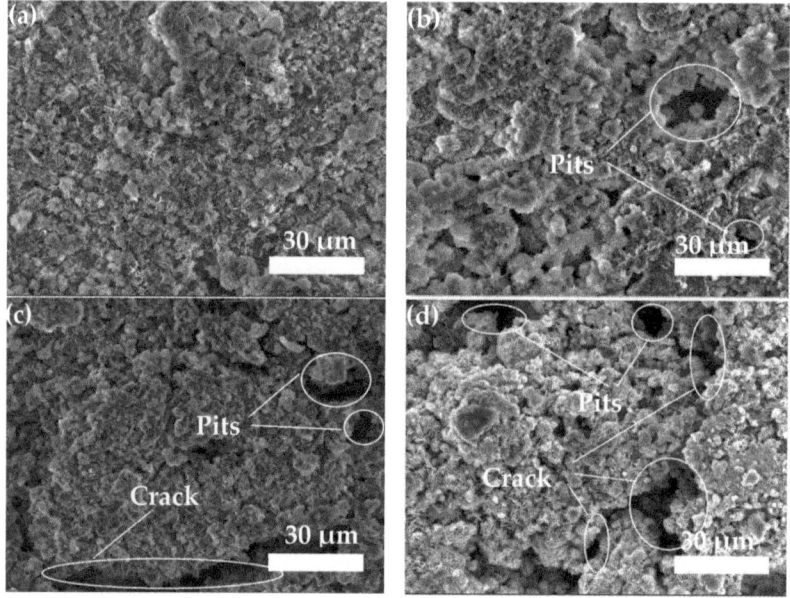

Figure 9. SEM micrographs for the $Zn_{65}Al_{15}Mg_5ZnO_{15}$ coating after different conditioning periods: (**a**) 120 h, (**b**) 240 h, (**c**) 360 h, and (**d**) 480 h.

Figure 10 is the composition analysis of corrosion products of $Zn_{65}Al_{15}Mg_5ZnO_{15}$ and $Zn_{45}Al_{35}Mg_5ZnO_{15}$ by X-ray diffraction. It shows that the corrosion products of the two composite coatings are basically similar. The main corrosion products are $(Al_2O_3)_4 \cdot H_2O$, $Zn_5(OH)_8Cl_2 \cdot H_2O$, and $Zn(OH)_2$. The main ion reaction is as follows:

$$\text{Cathodic reaction: } O_2 + 2H_2O + 4e^- \rightarrow 4OH^- \quad (2)$$

$$\text{Oxidation of metallic zinc: } Zn - 2e^- \rightarrow Zn^{2+} \quad (3)$$

$$\text{Hydrolysis of zinc ions: } Zn^{2+} + 2H_2O \rightarrow Zn(OH)_2 + 2H^+ \quad (4)$$

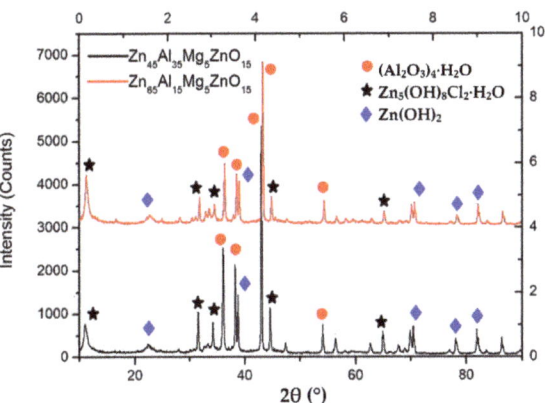

Figure 10. XRD analysis of the corrosion products of $Zn_{65}Al_{15}Mg_5ZnO_{15}$ and $Zn_{45}Al_{35}Mg_5ZnO_{15}$.

The corrosion rate of the two composite coatings becomes slower throughout the corrosion process, and then the corrosion rate becomes faster. Due to the corrosion rate of the coating being relatively fast at first, the corrosion of the coating produces surface oxidation of the oxidation product, which slows down the corrosion rate. When the passivation film is etched, the coating continues to be etched, so the corrosion rate is increased. The entire corrosion process is repeated, and it is known that the coating has been etched.

The corrosion resistance test data show that the $Zn_{65}Al_{15}Mg_5ZnO_{15}$ and $Zn_{45}Al_{35}Mg_5ZnO_{15}$ composite coatings both have an acceptable corrosion resistance but that the $Zn_{45}Al_{35}Mg_5ZnO_{15}$ composite coatings have a superior corrosion resistance and durability to the $Zn_{65}Al_{15}Mg_5ZnO_{15}$ composite coatings.

3.4. Photocatalytic Degradation of Methyl Blue

Table 2 and Figure 11 are the results of degradation of methyl blue solution by the $Zn_{65}Al_{15}Mg_5ZnO_{15}$ composite coating, the $Zn_{45}Al_{35}Mg_5ZnO_{15}$ composite coating, and the matrix steel Q235 under ultraviolet irradiation. Table 2 shows that the absorbance of methyl blue is the fastest under the action of the $Zn_{45}Al_{35}Mg_5ZnO_{15}$ composite coating and that the fading rate of methyl blue degradation solution is also the fastest. The absorbance of methyl blue is the slowest under the action of Q235, and the fading rate of methyl blue degradation solution is the slowest. The peak data of methyl blue absorbance were calculated to obtain the degradation rate of methyl blue. Figure 11 shows that the $Zn_{65}Al_{15}Mg_5ZnO_{15}$ composite coating and the $Zn_{45}Al_{35}Mg_5ZnO_{15}$ composite coating degrade methyl blue faster than Q235. In the same time, the degradation rate of the $Zn_{45}Al_{35}Mg_5ZnO_{15}$ composite coating to methyl blue was 98.9%, the degradation rate of methyl blue by the $Zn_{65}Al_{15}Mg_5ZnO_{15}$ composite coating was 96.7%, and the degradation rate of methyl blue by Q235 was only 71.8%.

Table 2. Degradation of methyl blue solution by $Zn_{65}Al_{15}Mg_5ZnO_{15}$, $Zn_{45}Al_{35}Mg_5ZnO_{15}$, and matrix steel Q235 under ultraviolet light.

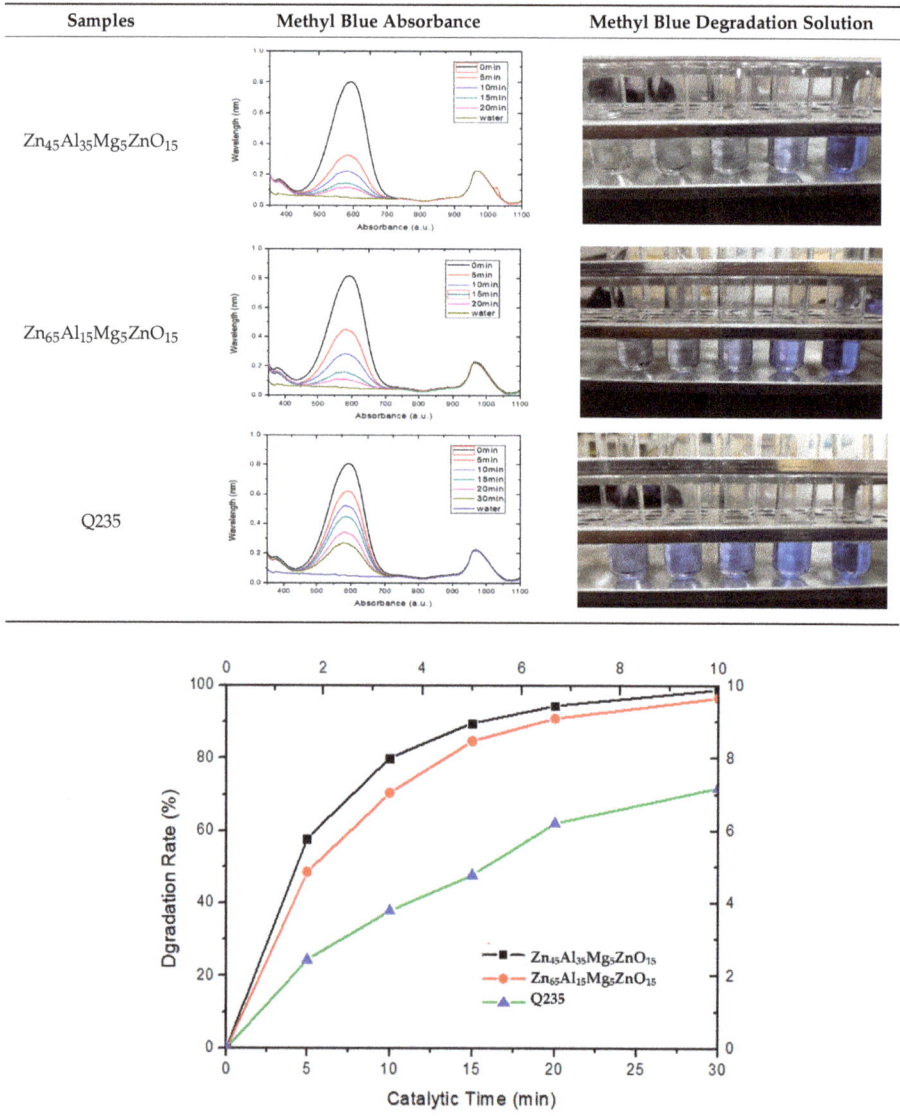

Figure 11. Degradation rate of equivalent concentration of methyl blue under ultraviolet light at different times for $Zn_{65}Al_{15}Mg_5ZnO_{15}$, $Zn_{45}Al_{35}Mg_5ZnO_{15}$, and base steel Q235.

The methyl blue degradation test data show that the degradation rates of methyl blue by the $Zn_{65}Al_{15}Mg_5ZnO_{15}$ composite coating and the $Zn_{45}Al_{35}Mg_5ZnO_{15}$ composite coating were much faster than that of Q235 and that the coatings have good photocatalytic degradation. The degradation efficiency of the $Zn_{45}Al_{35}Mg_5ZnO_{15}$ composite coating on methyl blue is better than that of the $Zn_{65}Al_{15}Mg_5ZnO_{15}$ composite coating.

3.5. Friction and Wear Test of Composite Coatings

Figure 12 shows the friction and wear of the $Zn_{65}Al_{15}Mg_5ZnO_{15}$ composite coating, the $Zn_{45}Al_{35}Mg_5ZnO_{15}$ composite coating, and the base steel Q235 under the same conditions. Figure 12a shows the appearance of the $Zn_{45}Al_{35}Mg_5ZnO_{15}$ composite coating after friction and wear, and the depth of the wear scar is 28 ± 1 µm. Figure 12b shows the appearance of the $Zn_{45}Al_{35}Mg_5ZnO_{15}$ composite coating after friction and wear, and the depth of the wear scar is 44 ± 1 µm. Figure 12c shows the appearance of the Q235 steel after friction and wear, and the depth of the wear scar is 180 ± 5 µm. It can be seen that the wear marks of the $Zn_{45}Al_{35}Mg_5ZnO_{15}$ composite coatings are the shallowest, followed by the $Zn_{65}Al_{15}Mg_5ZnO_{15}$ composite coatings, and the wear marks of the Q235 matrix steel are the deepest.

Table 3 shows the friction and wear data of $Zn_{65}Al_{15}Mg_5ZnO_{15}$, $Zn_{45}Al_{35}Mg_5ZnO_{15}$, and base steel Q235. Table 3 shows that the friction and wear coefficient of the $Zn_{45}Al_{35}Mg_5ZnO_{15}$ composite coating is 0.181, and the amount of wear is 0.0035 g. The friction and wear coefficient of the $Zn_{65}Al_{15}Mg_5ZnO_{15}$ composite coating is 0.231 and the wear amount is 0.0050 g. The base steel Q235 has a friction and wear coefficient of 0.358 and a wear amount of 0.0111 g. By comparing the three groups of experimental data, it can be seen that the friction coefficient and wear amount of the $Zn_{45}Al_{35}Mg_5ZnO_{15}$ composite coating are the smallest and that its wear resistance is the best.

Table 3. Friction and wear data of $Zn_{65}Al_{15}Mg_5ZnO_{15}$, $Zn_{45}Al_{35}Mg_5ZnO_{15}$, and base steel Q235.

Samples	m_1 (g)	m_2 (g)	Δm (g)	f
$Zn_{45}Al_{35}Mg_5ZnO_{15}$	2.6735	2.6700	0.0035	0.181
$Zn_{65}Al_{15}Mg_5ZnO_{15}$	2.6420	2.6370	0.0050	0.231
Q235	1.7734	1.7623	0.0111	0.358

Note: m_1 is the quality before wear, m_2 is the quality after wear, Δm is the amount of wear, and f is the coefficient of wet friction.

Figure 12 and Table 3 show that the $Zn_{65}Al_{15}Mg_5ZnO_{15}$ composite coating and the $Zn_{45}Al_{35}Mg_5ZnO_{15}$ composite coating have better friction and wear resistance than the base steel Q235, and the $Zn_{45}Al_{35}Mg_5ZnO_{15}$ composite coating has the best friction and wear characteristics.

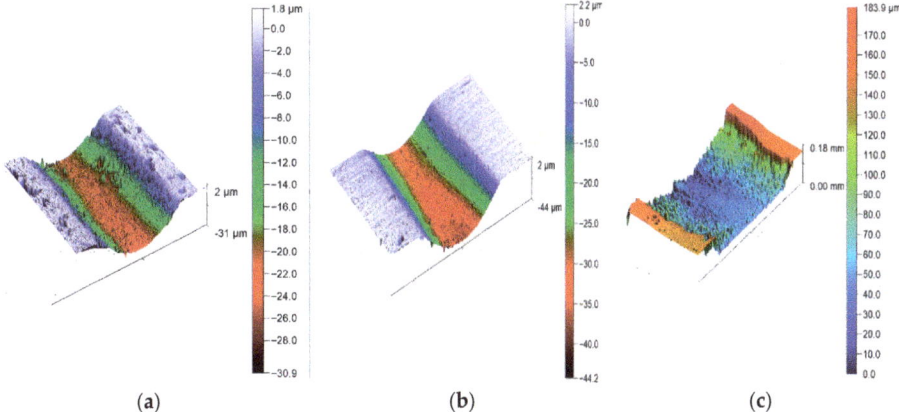

Figure 12. Friction and wear appearance of $Zn_{65}Al_{15}Mg_5ZnO_{15}$, $Zn_{45}Al_{35}Mg_5ZnO_{15}$, and Q235 matrix steel: (a) $Zn_{45}Al_{35}Mg_5ZnO_{15}$, (b) $Zn_{65}Al_{15}Mg_5ZnO_{15}$, and (c) Q235.

4. Conclusions

Two types of Zn–Al–Mg/ZnO composite coatings were successfully prepared on Q235 metallic substrate involving the cold spray technique. The obtained coatings were investigated from corrosion performance, photocatalytic degradation, and wear resistance viewpoints. The following conclusions can be drawn:

- $Zn_{65}Al_{15}Mg_5ZnO_{15}$ composite coating and $Zn_{45}Al_{35}Mg_5ZnO_{15}$ composite coating can be prepared on base steel Q235 by cold spray technique. The surface of the $Zn_{45}Al_{35}Mg_5ZnO_{15}$ composite coating is smoother and flatter than the $Zn_{65}Al_{15}Mg_5ZnO_{15}$ composite coating. The micromorphology of the two composite coatings prepared by cold spraying is dense, there are no defects such as voids inside the coating, and the Zn, Al, Mg, and ZnO elements inside the coating are uniformly distributed.
- The analysis of electrochemical workstation test data shows that both the $Zn_{65}Al_{15}Mg_5ZnO_{15}$ and $Zn_{45}Al_{35}Mg_5ZnO_{15}$ composite coatings have strong anodic oxidation cathodic protection characteristics, which can protect the matrix steel Q235 from corrosion. The open circuit potential and polarization current of the two composite coatings at different time periods show that the corrosion rate of the $Zn_{65}Al_{15}Mg_5ZnO_{15}$ composite coating is higher than that of the $Zn_{45}Al_{35}Mg_5ZnO_{15}$ composite coating, and the corrosion products produced by the oxidation of the composite coating have corrosion slowing characteristics during corrosion.
- The neutral salt spray test of the two composite coatings shows that the $Zn_{45}Al_{35}Mg_5ZnO_{15}$ composite coating experienced regular delamination corrosion with no pits or cracks defects but the $Zn_{65}Al_{15}Mg_5ZnO_{15}$ composite coating was unevenly corroded and had a deep corrosion, with defects such as pits and cracks, after 240 h of corrosion. The corrosion resistance test shows that the $Zn_{45}Al_{35}Mg_5ZnO_{15}$ composite coating has a better corrosion resistance and long-term stability than the $Zn_{65}Al_{15}Mg_5ZnO_{15}$ composite coating.
- By comparing the photocatalytic degradation data of the $Zn_{65}Al_{15}Mg_5ZnO_{15}$ composite coatings, $Zn_{45}Al_{35}Mg_5ZnO_{15}$ composite coatings, and Q235 matrix steel for methyl blue solution, the results show that the two composite coatings have good photocatalytic properties but that the photocatalytic properties of the $Zn_{45}Al_{35}Mg_5ZnO_{15}$ composite coatings are better.
- The results of the friction and wear tests show that both the $Zn_{65}Al_{15}Mg_5ZnO_{15}$ composite coating and the $Zn_{45}Al_{35}Mg_5ZnO_{15}$ composite coating have better friction and wear resistance than the base steel Q235 and that the friction and wear properties of the $Zn_{45}Al_{35}Mg_5ZnO_{15}$ composite coating are the best.
- The results of all the above experiments show that the $Zn_{45}Al_{35}Mg_5ZnO_{15}$ composite coating prepared by cold spraying has a superior deposition effect, high photocatalytic efficiency, better corrosion resistance, and wear resistance.

Author Contributions: Conceptualization, S.W.; methodology, X.L. and S.W.; validation, X.L. and G.W.; data curation, S.W. and X.L.; writing—original draft preparation, X.L.; writing—review and editing, S.W. and D.W.; supervision, T.X. and H.D.; project administration, S.W.

Funding: This research was funded by the National Natural Science Foundation of China (No. 51872122), the Shandong Key Research and Development Plan, China (No.: 2016JMRH0218), and the Taishan Scholar Engineering Special Funding (2016–2020).

Conflicts of Interest: The authors declare no conflict of interest.

References

1. Yan, S.K.; Song, G.L.; Li, Z.X.; Wang, H.N.; Zheng, D.J.; Cao, F.Y.; Horynova, H.; Dargusch, M.S.; Zhou, L. A state-of-the-art review on passivation and biofouling of Ti and its alloys in marine environments. *J. Mater. Sci. Technol.* **2018**, *34*, 421–435. [CrossRef]
2. Chavan, N.M.; Kiran, B.; Jyothirmayi, A.; Phani, P.S.; Sundararajan, G. The corrosion behavior of cold sprayed zinc coatings on mild steel substrate. *J. Therm. Spray Technol.* **2013**, *22*, 463–470. [CrossRef]
3. Olakanmi, E.O.; Doyoyo, M. Laser-assisted cold-sprayed corrosion-and wear-resistant coatings: A review. *J. Therm. Spray Technol.* **2014**, *23*, 765–785. [CrossRef]

4. Cai, M.R.; Guo, R.S.; Zhou, F.; Liu, W.M. Lubricating a bright future: Lubrication contribution to energy saving and low carbon emission. *Sci. China Technol. Sci.* **2013**, *56*, 2888–2913. [CrossRef]
5. Cui, H.; Li, N.; Peng, J.; Yin, R.; Li, J.; Wu, Z. Investigation on the thermal performance of a novel spray tower with upward spraying and downward gas flow. *Appl. Energy* **2018**, *231*, 12–21. [CrossRef]
6. Gkomoza, P.; Lampropoulos, G.S.; Vardavoulias, M.; Pantelis, D.I.; Karakizis, P.N.; Sarafoglou, C. Microstructural investigation of porous titanium coatings, produced by thermal spraying techniques, using plasma atomization and hydride-dehydride powders, for orthopedic implants. *Surf. Coat. Technol.* **2019**, *357*, 947–956. [CrossRef]
7. Wu, Y.N.; Ke, P.L.; Wang, Q.M.; Sun, C.; Wang, F.H. High temperature properties of thermal barrier coatings obtained by detonation spraying. *Corros. Sci.* **2004**, *46*, 2925–2935. [CrossRef]
8. Pawłowski, L. Strategic oxides for thermal spraying: Problems of availability and evolution of prices. *Surf. Coat. Technol.* **2013**, *220*, 14–19. [CrossRef]
9. Xie, X.; Chen, C.; Xie, Y.; Aubry, E.; Ren, Z.; Ji, G.; Liao, H. Comparative investigation of microstructure and properties of Ni-coated FeSiAl soft magnetic composite coatings produced by cold spraying and HVOF. *Surf. Coat. Technol.* **2019**, *371*, 224–234. [CrossRef]
10. Hunter, B.; Aldwell, B.; Jenkins, R.; Lupoi, R. A study on the feasibility of laser annealing to relieve residual stresses in cold spray coatings. *Procedia CIRP* **2018**, *78*, 91–96. [CrossRef]
11. Xie, Y.C.; Chen, C.Y.; Planche, M.P.; Deng, S.H.; Huang, R.Z.; Ren, Z.M.; Liao, H.L. Strengthened peening effect on metallurgical bonding formation in cold spray additive manufacturing. *J. Therm. Spray Technol.* **2019**, *28*, 769–779. [CrossRef]
12. Chen, C.Y.; Xie, Y.C.; Yin, Y.S.; Fukanuma, H.; Huang, R.Z.; Zhao, R.X.; Wang, J.; Ren, Z.G.; Liu, M.; Liao, H.L. Effect of hot isostatic pressing (HIP) on microstructure and mechanical properties of Ti6Al4V alloy fabricated by cold spray additive manufacturing. *Addit. Manuf.* **2019**, *27*, 595–605. [CrossRef]
13. Kang, N.; Coddet, P.; Liao, H.L.; Coddet, C. The effect of heat treatment on microstructure and tensile properties of cold spray Zr base metal glass/Cu composite. *Surf. Coat. Technol.* **2015**, *280*, 64–71. [CrossRef]
14. Coddet, P.; Verdy, C.; Coddet, C.; Debray, F. Effect of cold work, second phase precipitation and heat treatments on the mechanical properties of copper–silver alloys manufactured by cold spray. *Mater. Sci. Eng. A* **2015**, *637*, 40–47. [CrossRef]
15. Esmaily, M.; Svensson, J.E.; Fajardo, S.; Birbilis, N.; Frankel, G.S.; Virtanen, S.; Arrabal, R.; Thomas, S.; Johansson, L.G. Fundamentals and advances in magnesium alloy corrosion. *Prog. Mater. Sci.* **2017**, *89*, 92–193. [CrossRef]
16. Duchoslav, J.; Steinberger, R.; Arndt, M.; Keppert, T.; Luckeneder, G.; Stellnberger, K.H.; Hagler, J.; Angeli, G.; Riener, C.K.; Stifter, D. Evolution of the surface chemistry of hot dip galvanized Zn–Mg–Al and Zn coatings on steel during short term exposure to sodium chloride containing environments. *Corros. Sci.* **2015**, *91*, 311–320. [CrossRef]
17. Grégoire, B.; Bonnet, G.; Pedraza, F. Mechanisms of formation of slurry aluminide coatings from Al and Cr microparticles. *Surf. Coat. Technol.* **2019**, *359*, 323–333. [CrossRef]
18. Li, C.G.; Li, S.; Zeng, M.; Sun, S.; Wang, F.F.; Wang, Y. Effect of high-frequency micro-vibration on microstructure and properties of laser cladding aluminum coatings. *Int. J. Adv. Manuf. Technol.* **2019**, *103*, 1633–1642. [CrossRef]
19. La, P.; Ou, Y.; Han, S.; Wei, Y.; Zhu, D.; Feng, J. Effect of carbon content on morphology, size and phase of submicron tungsten carbide powders by salt-assisted combustion synthesis. *Rare Metal Mater. Eng.* **2016**, *45*, 853–857. [CrossRef]
20. Li, H.X.; Li, X.B.; Sun, M.-X.; Wang, H.R.; Huang, G.S. Corrosion resistance of cold-sprayed Zn–50Al coatings in seawater. *J. Chin. Soc. Corros. Prot.* **2010**, *30*, 62–66. [CrossRef]
21. Liang, Y.L.; Wang, Z.B.; Zhang, J.; Zhang, J.B.; Lu, K. Enhanced bonding property of cold-sprayed Zn–Al coating on interstitial-free steel substrate with a nanostructured surface layer. *Appl. Sur. Sci.* **2016**, *385*, 341–348. [CrossRef]
22. Kumar, V.A.; Sammaiah, P. Comparison of process parameters influence on mechanical and metallurgical properties of zinc coating on mild steel and aluminium during mechanical process. *Mater. Today Proc.* **2018**, *5*, 3861–3866. [CrossRef]

23. Mandal, G.K.; Das, S.K.; Balasubramaniam, R.; Mehrotra, S.P. Evolution of microstructures of galvanised and galvannealed coatings formed in 0.2 wt.% aluminium–zinc bath. *Mater. Sci. Technol.* **2011**, *27*, 1265–1270. [CrossRef]
24. Darvish, A.; Naderi, R.; Attar, M.R.M. Improvement in polyurethane coating performance through zinc aluminium phosphate pigment. *Pigment. Resin Technol.* **2016**, *45*, 419–425. [CrossRef]
25. La, J.H.; Lee, S.Y.; Hong, S.J. Synthesis of Zn–Mg coatings using unbalanced magnetron sputtering and theirs corrosion resistance. *Surf. Coat. Technol.* **2014**, *259*, 56–61. [CrossRef]
26. Zhang, X.Y.; Shi, Z.C.; Guo, M.Q.; Lu, F.; Sun, Z.H.; Tang, Z.Z.; Yu, B. Morphology and properties of cold sprayed Zn–Ni composite coatings. *Therm. Spray. Technol.* **2014**, *6*, 6–13. (In Chinese) [CrossRef]
27. Conde, A.; Arenas, M.A.; De Damborenea, J.J. Electrodeposition of Zn–Ni coatings as Cd replacement for corrosion protection of high strength steel. *Corros. Sci.* **2011**, *534*, 1489–1497. [CrossRef]
28. Gayathri, P.V.; Yesodharan, S.; Yesodharan, E.P. Microwave/Persulphate assisted ZnO mediated photocatalysis (MW/PS/UV/ZnO) as an efficient advanced oxidation process for the removal of RhB dye pollutant from water. *J. Environ. Chem. Eng.* **2019**, *7*, 103122. [CrossRef]
29. Singh, N.K.; Koutu, V.; Malik, M.M. Enhancement of room temperature ferromagnetic behavior of Co-doped ZnO nanoparticles synthesized via sol–gel technique. *J. Sol-Gel Sci. Technol.* **2019**, *91*, 324–334. [CrossRef]
30. Goel, S.; Kumar, B. Lead-free high Tc ferroelectric material: Hierarchical Dy-doped ZnO architectures co-assembled by 1D nanorods and 2D nanosheets. *J. Alloy. Compd.* **2019**, *801*, 626–639. [CrossRef]

© 2019 by the authors. Licensee MDPI, Basel, Switzerland. This article is an open access article distributed under the terms and conditions of the Creative Commons Attribution (CC BY) license (http://creativecommons.org/licenses/by/4.0/).

Article

The Abrasive Wear Resistance of Coatings Manufactured on High-Strength Low-Alloy (HSLA) Offshore Steel in Wet Welding Conditions

Jacek Tomków [1,*], Artur Czupryński [2] and Dariusz Fydrych [1]

[1] Faculty of Mechanical Engineering, Gdańsk University of Technology, G. Narutowicza 11/12, 80–233 Gdańsk, Poland; dariusz.fydrych@pg.edu.pl
[2] Department of Welding Engineering, Silesian University of Technology, Konarskiego 18A, 44–100 Gliwice, Poland; artur.czuprynski@polsl.pl
* Correspondence: jacek.tomkow@pg.edu.pl; Tel.: +48-58-347-1863

Received: 3 February 2020; Accepted: 27 February 2020; Published: 29 February 2020

Abstract: Some marine and offshore structure elements exploited in the water cannot be brought to the surface of the water as this will generate high costs, and for this reason, they require in-situ repairs. One of the repair techniques used in underwater pad welding conditions is a wet welding method. This paper presents an investigation of the abrasive wear resistance of coatings made in wet welding conditions with the use of two grades of covered electrodes—an electrode for underwater welding and a commercial general use electrode. Both electrodes were also used for manufacturing coatings in the air, which has been also tested. The Vickers HV10 hardness measurements are performed to demonstrate the correlation in abrasive wear resistance and the hardness of each specimen. The microscopic testing was performed. For both filler materials, the coatings prepared in a water environment are characterized by higher resistance to metal–mineral abrasion than coatings prepared in an air environment—0.61 vs. 0.44 for commercial usage electrode and 0.67 vs. 0.60 for underwater welding. We also proved that in the water, the abrasive wear was greater for specimens welded by the general use electrode, which results in a higher hardness of the layer surface. In the air welding conditions, the layer welded by the electrode for use in the water was characterized by a lower hardness and higher resistance to metal–mineral abrasion. The microstructure of the prepared layers is different for both the environment and both electrodes, which results in abrasive wear resistance.

Keywords: wet welding; underwater welding; abrasive wear resistance; high-strength low-alloy steel; hardness measurements; metal–mineral abrasion

1. Introduction

The number of offshore structures that are operated in changing and harsh conditions increases each year [1]. The aggressive environmental factors are responsible for offshore damages, due to, e.g., fatigue, mechanical damages, erosion and corrosion [2–4]. Undersea structures can be damaged due to abrasion as a consequence of the movement of sand and other solid particles caused by sea currents and waves. In addition, the internal surfaces of fluid transport pipelines may be damaged due to abrasion and cavitation [5]. Apart from typical damage in the form of cracks, these phenomena can cause a reduction in the cross-section of the elements. These factors make it necessary to use protective coatings on the steel surfaces, which improve the corrosion and wear resistance [6,7].

These coatings can be deposited by thermal spraying during the production of offshore structures [8]. In some applications, the composite coatings can be used [9,10]. However, the most commonly deposited layers are manufactured by the use of ceramics and metals. The ceramic coatings

Al$_2$O$_3$/TiO$_2$, ZrO$_2$/CaO can be applied by a thermal spraying process [11–13]. The arc thermal spraying is also used to deposit the metals and its alloys [14,15]. The metals can be applied by laser cladding [16], arc welding [17] and friction surfacing [18]. Information on thermal spraying directly in water is very limited. Liu et al. proposed a novel underwater repairing and remanufacturing process and characterized the impact of the water environment on the process of applying aluminum coating by the cold spray method on a steel substrate [19]. Another promising technology to repair damaged components and prepare coatings is in-situ underwater laser welding [20–22]. However, most of the coatings used in offshore structures are applied by arc welding processes during manufacturing [23,24].

The research works are focused on the improvement of wear and corrosion resistance of laid layers. One of the methods to reduce the fracture of coatings could be the use of heating/cooling controls and buffers [25]. The coating applications are made in an air environment. Increasing the interpass temperature decreases the abrasive wear resistance, which is a result of the increase in the grain size [26]. Unfortunately, the offshore structures still need to be repaired, which must be carried out in the water environment (in-situ) as it is lower in cost than in the case of transferring the structures into air welding conditions. The most often used underwater repair method underwater is wet welding. This process is carried out in direct contact with water and can be performed using different methods, e.g., by the self-shielded flux-cored arc welding (FCAW-S) method [27–29]. The FCAW-S process does not require the use of shielding gas, and it is easy to automate and train welders [23,27–30]. However, manual metal arc (MMA) welding is still the most commonly used as an underwater welding process [31,32].

The water environment generates some essential problems during repair works. The first is a higher diffusible hydrogen content in deposited metal than in air conditions, which is the result of the dissociation of water in the welding environment [32,33]. The problems of hydrogen embrittlement increase when the structure is in sea water [34]. The second disadvantage associated with welding underwater is the high susceptibility of steel to form brittle structures in the heat-affected zone (HAZ), which causes high values of residual stress [35]. These factors, including a high cooling rate, are responsible for the cold cracking of high-strength low-alloy steel (HSLA) joints made in underwater conditions [35,36].

As the results of wear tests of the underwater welded joints properties are extremely rarely reported, in this study, underwater manual metal arc coating deposition on HSLA S355G10+N steel was carried out. The deposit appearance, microstructure and hardness of specimens were investigated and compared with the results of tests carried out on specimens welded in an air environment. In addition, tests on abrasion resistance were carried out for specimens manufactured in both conditions. To the best of the author's knowledge, in the literature, there are no researches published that provide a detailed discussion on the problem of abrasive wear resistance of coatings manufactured in the water environment.

2. Materials and Methods

2.1. Used Materials

For welding, S355G10+N steel plates (U.S. STEEL, Košice, Slovak Republic) with dimensions of 120 × 100 × 16 mm were chosen as a base material (BM). This is an HSLA steel grade dedicated to making offshore structure elements [9]. The chemical composition of the BM has been tested by emission spectroscopy with a spark excitation method. As a filler material, two grades of covered electrodes were chosen. The first was an ISO 2560-A: E38 0 R11 [37] Omnia rutile electrode with a diameter of 4.0 mm for general use, which provides welds with good plasticity, which decreases the susceptibility to cold cracking. The second electrode was a specially formulated wet welding Barracuda Gold electrode—the nearest equivalent is ISO 2560-A: E42 2 1Ni RR 51 [37]. The chemical compositions of the used materials are listed in Table 1.

Table 1. Chemical compositions of used materials wt %.

Material	C	Si	Mn	P	Cr	Mo	Ni	Al	Cu	S	Ce$_{IIW}$ **
S355G10+N	0.11	0.35	1.39	0.008	0.02	0.02	0.25	0.039	0.27	-	0.385
Omnia E38 0 R 11 electrodes deposit *	0.07	0.45	0.50	0.010	-	-	-	-	0.05	-	-
Barracuda Gold E42 2 1Ni RR 51 deposit *	0.05	0.45	0.55	0.025	-	-	0.30	-	-	0.025	-

* In accordance to manufacturer data; **Ce$_{IIW}$—carbon equivalent by International Institute of Welding.

2.2. Welding Procedure

For the tests, four specimens were prepared with the use of MMA. Two of them were welded in the water environment (0.25 m water depth) by two different grades of covered electrodes. Two specimens were made in the air by different electrodes. To cover all surfaces of the steel plate, the number of beads were different for each environment: for specimens welded in the air—12 beads—and for specimens made underwater—14 beads. The welding parameters were chosen according to previous investigations [32,35,38] and the manufacturer's data. Welding with Omnia electrodes was carried out by a welding current with the range 150–160 A (the manufacturer allows the range 130–160 A), and with Barracuda Gold electrodes the range 190–200 A (the manufacturer allows the range 170–220 A). These parameters produced heat input values in the range of 0.7–0.8 kJ/mm for Omnia electrodes and higher than 0.9 kJ/mm for Barracuda Gold electrodes, which limits the susceptibility to cold cracking of used steel. For specimens made in the air, beads were laid one after the other. The research plan developed at the beginning assumed conducting the experiment in conditions as close as possible to real conditions, used in industrial applications. During welding in the air of the tested steel grade, there is usually no need to use a special bead sequence. However, during welding in the water environment, due to the need to reduce the tendency of steel to cold cracking, a change in welding technique is recommended, which was found in our previous study [38]. Specimens welded underwater were characterized by a lower width of laid beads. This resulted in a higher number of beads for underwater welding conditions than for specimens manufactured in the air. The scheme of the specimens is presented in Figure 1.

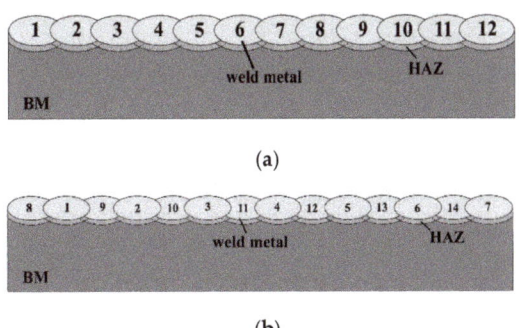

Figure 1. The schematic view of the specimens with the order of laid beads (a) welded in the air and (b) welded underwater.

2.3. Examination Procedure

After welding, steel plates were cut for specimens for hardness HV10 measurements (one specimen from each plate) and for abrasive wear resistance tests (two specimens from each plate). For specimens, for the abrasive wear test, after cutting, the upper surfaces of specimens were groundwith a surface grinder to ensure parallelism with the lower surface. The scheme of cutting a steel plate is presented in Figure 2.

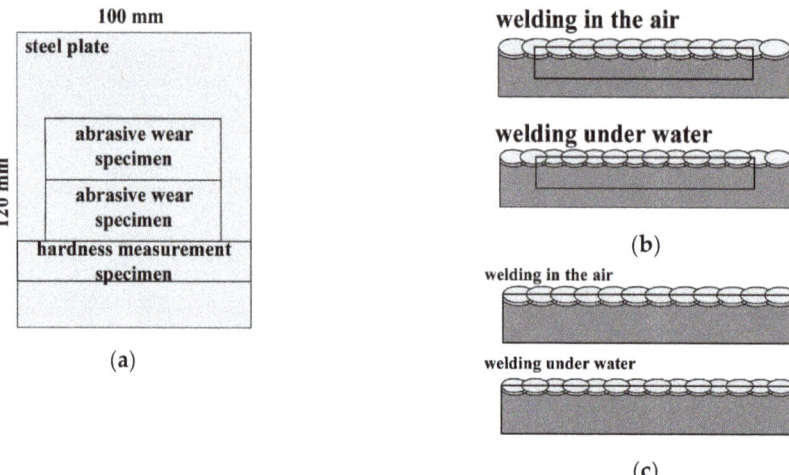

Figure 2. The schematic view of cutting of specimens for investigations: (**a**) top view; (**b**) cross-section—specimen for abrasive wear resistance test; (**c**) cross-section—specimen for hardness measurements.

To determine quantitatively the abrasive wear resistance of Omnia and Barracuda Gold electrode layer MMA surface deposits, the tests of abrasive wear type metal–ceramic were conducted in accordance with standard ASTM G 65-00 Procedure A [39]. The 25 mm wide and 75 mm long abrasive wear resistance test specimens were cut from the middle area of Omnia and Barracuda Gold electrode layer MMA surface deposits. All specimens were weighed to the nearest 0.0001 g, as required by ASTM G65-00. During the tests, the rubber wheel apparatus made six thousand revolutions, and the flow rate of the abrasive (sand A. F. S. Testing Stand 50–70) was 335 g/min. The force applied to press the test coupon against the wheel was TL = 130 N (test load (TL)). After the abrasive wear resistance test, the specimen was weighed at a weight sensitivity of 0.0001 g. The mass loss of specimens of Omnia and Barracuda Gold electrode layer MMA surface deposits was reported directly and relatively in comparison to the mass loss of the reference Hardox 450 steel, which is commonly used [40]. Next, the densities of surface deposits were measured and the abrasive tests' results were reported. The deposits density was determined using hydrostatic RADWAG WAX 60/220 weight, based on the three measurements of specimens taken from each of the prepared coatings weighed in the air and in the water. The schematic diagram of ASTM G65-00 Procedure A abrasive wear resistance test and test stand are presented in Figure 3.

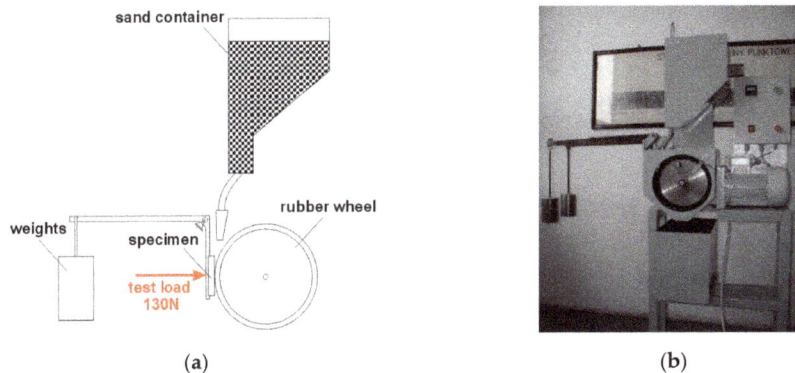

Figure 3. Schematic diagram of ASTM G65-00 Procedure A abrasive wear resistance test (**a**) and apparatus overview (**b**).

The Vickers HV10 hardness measurements were taken in accordance with EN ISO 9015:2011 [41], using a Sinowon V-10 stand, with a measurement error of ±3HV10. Before measurements were performed, the tested cross-sections were ground by sandpaper with a granulation of up to 1000. Then, they were etched by Nital 4% to show the structure of the layer and HAZ and to calculate the depth that the upper surface was ground to. Then, the upper surfaces wereground at the surface grinder to ensure the parallelism with the lower surface. The hardness HV10 measurements were prepared on the upper surface of the specimen and on cross-sections in accordance with the two measurement layers. The schematic view of the hardness HV10 measurements is presented in Figure 4.

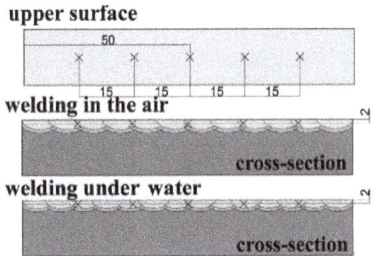

Figure 4. The scheme of the hardness measurement points.

The microscopic tests were done in accordance with the EN ISO 17639:2013 [42]. Observations were carried out with the use of an optical microscope. From each specimen, the welds and HAZ were observed. The specimens wereground, polished and etched by Nital 4%.

3. Results and Discussions

3.1. Abrasive Wear Resistance

The density of Omnia and Barracuda Gold electrode layer MMA surface deposits were measured and the abrasive tests' results were reported as volume loss in cubic millimeters, by converting mass loss to volume loss as follows [39]:

$$\text{Volume loss, (mm}^3\text{)} = \text{mass loss (g)} : \text{density (g/cm}^3\text{)} \times 1000, \qquad (1)$$

The results of low-stress abrasion wear resistance to the metal–ceramic scratching when using dry quartz sand as the abrasion material of the HARDOX 450 wear plate and MMA surface layer deposits of the Omnia and Barracuda Gold electrodes tested are presented in Table 2.

Table 2. Results of abrasive wear resistance test.

Specimen Designation	Number of Specimens	Weight before Test (g)	Weight after Test (g)	Mass Loss (g)	Density (g/cm^3)	Volume Loss (mm^3)	Average [1] Volume Loss (mm^3)	Relative [2] Abrasion Resistance
O (OMNIA 46 electrode)	W1	156.6923	155.4087	1.2836	7.6927	166.8595	174.7537(11.1641)	0.61
	W2	157.5938	156.1770	1.4168	7.7570	182.6479		
	A1	152.3573	150.9207	1.4366	6.9626	206.3310	240.0881(47.7398)	0.44
	A2	158.2508	156.1954	2.0554	7.5057	273.8452		
B (Barracuda Gold electrode)	W1	152.0878	150.8147	1.2731	7.6565	166.2770	158.7265(10.6780)	0.67
	W2	148.5804	147.4337	1.1467	7.5852	151.1760		
	A1	151.3845	149.9140	1.4705	7.5319	195.2363	179.0374(22.9087)	0.60
	A2	144.0795	142.8275	1.252	7.6886	162.8385		
				Reference material				
H (Hardox 450)	1	112.5135	111.6854	0.8281	7.7531	106.8089	106.5779(0.3267)	1.0
	2	114.7009	113.8025	0.8284	7.7896	106.3469		

Remarks: [1] Standard deviation in brackets; [2] relative to Hardox 450 steel plate. Test load on specimen 130 N. W—specimen surfacing under water. A—specimen surfacing in the air atmosphere.

The views of the specimen abrasion marks after the test are presented in Figure 5. The observed surfaces of investigated coatings (Figure 5a–d) have typical morphology for abrasive wear resistance test. It was stated that two mechanisms are responsible for the wear of the surface. In addition to the expected rolling mechanism [43], the effect of the grooving mechanism was found. This effect results from sharp-edge abrasive particles, which penetrated deep into the investigated surface. The effect of penetration is a larger amount of material removal from specimens, which initiated the grooving mechanism [44]. In contrast to the wear mechanism for Hardox 450 steel (grooving), the rolling mechanism was predominant in investigated coatings. Another aspect of determining higher wear resistance of Hardox 450 (C=0.26%, Cr=1.40%) steel is lower carbon and chromium content in investigated coatings (Table 1). For this reason, the layers are intensively strained and cut by abrasive grains [45].

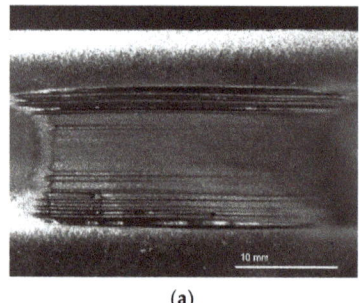
(a)

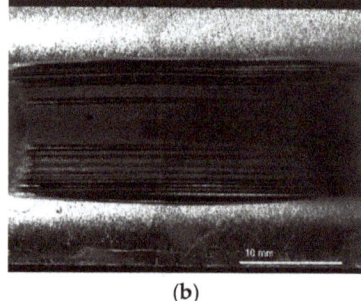
(b)

Figure 5. Cont.

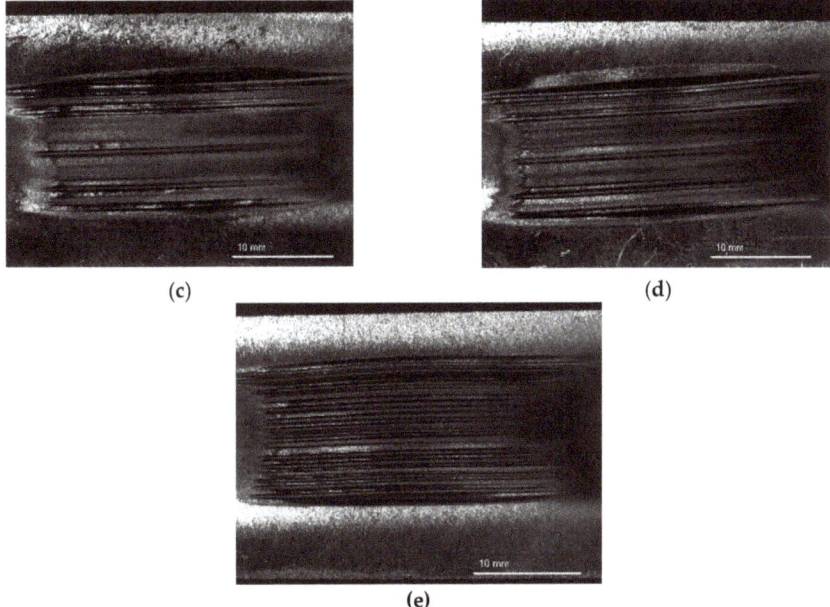

Figure 5. View of the specimen abrasion marks after the abrasive wear resistance test. O—Omnia Electrode; B—Barracuda Gold electrode; H—Hardox 450 steel; W—welding underwater; A—welding in the air; (**a**) OW1; (**b**) OA1; (**c**) BW1; (**d**) BA1; (**e**) H1.

Hardox 450 steel (samples H1, H2) had the greatest relative resistance to the metal–mineral abrasion. The surface of the arc hardfacing layers at the place of impact of the abrasive compared to Hardox 450 was characterized by a deeper metal wiping. The top layers padded with a covered electrode—Barracuda Gold—made both in a water environment and in the outdoor welding, compared to layers made in analogous conditions with an Omnia rutile electrode, are characterized by greater resistance to the metal–mineral abrasion, Figure 6. It should be emphasized here that in the case of both consumables used, the resistance to abrasive wear of layers made in a water environment was higher than that of layers made under normal conditions.

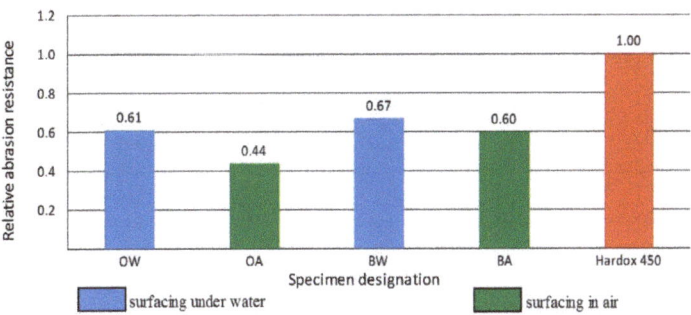

Figure 6. Comparison of the relative abrasion resistance of surfaced surface layers with the reference material—Hardox 450 steel.

3.2. Hardness HV10 Measurements

Before measurements were carried out, the macroscopic tests were performed for each cross-section in accordance with the EN ISO 17639:2013 [42]. There were no imperfections detected during macroscopic testing. An example view of the prepared cross-section is presented in Figure 7.

Figure 7. An example view of a cross-section of the layer made by an Omnia electrode in the air, with etching using Nital 4%.

The macroscopic photographs were used to calculate the dilution for layers performed in the air and in the water. The dilution is a factor that complicates predicting the effects of multi-pass welding, which affects, among other aspects, hardenability [46]. The dilution was calculated i.e., D = A_{base}/A, where A is the total molten area, and A_{base} is the molten area of BM [47]. The scheme of dilution calculation is presented in Figure 8.

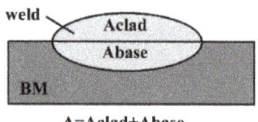

A=Aclad+Abase

Figure 8. The schematic view of the methodology of dilution calculations; A—total molten area, A_{clad}—molten cladded area and A_{base}—molten area of BM.

The dilution values for specimens welded in the air were higher than those for specimens welded underwater. These values depend on the bead sequences used for the surface of specimens, which was described in our previous studies [38]. Coronado et al. [48] stated that higher dilution levels reduced wear resistance. The different dilution values result from an assumed methodology of bead sequences.

The hardness HV10 measurements show differences in all specimens. For the specimens welded inthe air, the coating made by electrodes for underwater welding was characterized by higher hardness. Different situations were observedfor measurements of specimens welded in the water environment. The coating produced by electrodes for underwater welding was characterized by a lower hardness. The investigated S355G10+N steel is classified as material from group 2.1. in accordance with EN ISO 15614-1:2017 [49]. The maximum hardness assumed by this standard for this group cannot exceed that of 380HV10. The hardness measurement showed that the coatings made underwater by the Omnia electrode did not fulfill the requirements of hardness by the standard EN ISO 15614-1:2017. The results of the hardness HV10 measurements are presented in Table 3.

Table 3. The results of the hardness HV10 measurements.

Specimen	Place of Measurements	>Hardness HV10 (±3)				
		Measurement Point				
		1	2	3	4	5
OA	Upper surface	220	245	236	220	215
	Cross-section	218	234	242	230	228
BA	Upper surface	253	246	265	277	245
	Cross-section	242	253	270	242	252
OW	Upper surface	455	459	445	472	444
	Cross-section	441	451	462	445	439
BW	Upper surface	331	321	335	310	296
	Cross-section	342	333	327	330	337

After welding in the air, the coating made by the electrode for underwater welding was characterized by higher hardness and higher relative abrasion resistance than the coating welded by the Omnia electrode. In the case of underwater welding, the layer (Barracuda Gold electrode) with lower hardness was characterized by a higher relative abrasion resistance. The literature [48] showed that there is no relationship between the hardness and relative abrasion resistance, which was proven in this investigation. For both environments, the Barracuda Gold electrodes produced the layer with the higher wear resistance. The Barracuda Gold layers were welded with higher values of heat inputs, which decreased the cooling rate. This providesa higher abrasive wear resistance for layers produced by the Barracuda Gold electrodes. The same factor was responsible for higher abrasive wear resistance of layers welded inthe air. The water, as a welding environment, increased the cooling rate, which generated brittle structures [27–31,35,38] with lower wear resistance.

3.3. Microscopic Study

The microscopic observations confirmed the earlier results. There were observed significant differences in the weld and in HAZ of specimens produced by different electrodes in different environments. The layers prepared in the water are characterized by the presence of dendritic structure in welds, which consisted of bright fine-grained pearlite arranged in columns. At the boundaries of dendrites, the acicular ferrite grew from these columns (Figure 9a,b). In the welds performed by the Omnia electrodes, cracks were observed (Figure 9b). This may be the result of Ni content in the Barracuda Gold electrode deposit, which improved the mechanical properties of weld metal [50]. The structure of the welds of coatings performed in the air is different. They are characterized by a partial disappearance of the dendritic structure (Figure 9c,d). This area is built by a fine-grained ferrite and pearlite structure, which results from the lower cooling rate in the air in comparison to the water. The weld produced by Barracuda Gold in the air is characterized by finer grain size, which improved wear resistance. This is in agreement with the results presented by Cui et al. [51]. The structure of HAZ was similar for coatings performed by both electrodes in the air. The structure indicated the presence of refined and tempered low-carbon martensite mixed with normalization structures with fine ferrite and pearlite (Figure 9d). The HAZ of underwater specimens is characterized by the presence of brittle bainitic and martensitic structures, which results in cracking observed for both used electrodes. In specimens prepared by Barracuda Gold, these cracks are located perpendicular to the fusion line (Figure 9e). The cracks observed in specimens performed bythe Omnia electrodes are located perpendicular and parallel to the fusion line. The length of those cracks was much higher. It was stated in the literature [44] that the presence of the cracks influencesa lower abrasive wear resistance of performed coatings. The exemplary results of metallographic microscopic tests are presented in Figure 9.

(a) (b)

Figure 9. Cont.

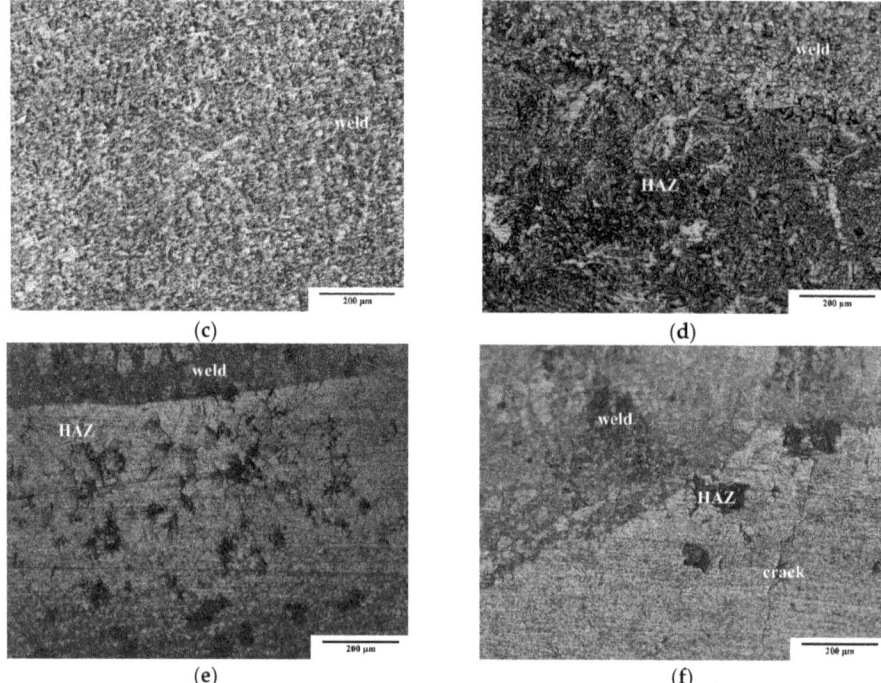

Figure 9. Exemplary results of microscopic testing:(**a**) weld made in the water by the Omnia electrode; (**b**) weld made in the water by the Barracuda Gold electrode;(**c**) weld made in the air by the Omnia electrode; (**d**) weld and heat-affected zone (HAZ) made in the air by the Barracuda Gold electrode; (**e**) HAZ of specimen made in the water—Barracuda Gold electrode; (**f**) HAZ of specimen made in the water—Omnia electrode.

4. Conclusions

The results of the performed investigations showed a significant influence of the welding environment on the hardness and wear resistance of the produced layers. We proved that relative abrasion resistance depends on the type of electrodes and the environment of welding. From a practical point of view, the most important finding is that there is no need to transport elements to the surface, and they can be repaired in-situ in the water. It will reduce the cost of maintaining and repairing of offshore infrastructure.

The performed examinations allow us to draw these conclusions:

1. Water, as an environment of the welding processes, generates higher HV10 hardness of steel coatings than air does. The microstructure of coatings performed in the water contains the brittle structures in the HAZ, which resulted from a high cooling rate. The structures for specimens manufactured in the air are characterized by the presence of normalized structures.
2. The hardness strongly depends on the welding environment, but no influence on the abrasive wear resistancewas found, neither in the air nor in the water environment.
3. The coatings produced by the electrodes for underwater welding (Barracuda Gold) are characterized by a lower volume loss in the metal–mineral abrasion conditions than layers welded by the general usage electrodes (Omnia). This is due to the smaller grain size of the weld metal produced in the air (Barracuda Gold), and the greater number of cracks for coating made in the water (Omnia).

4. The water environment of welding produces coatings with higher abrasive wear resistance than layers welded in the air, which resulted from different metallographic microstructures formed in coatings manufactured in other environments.

Author Contributions: Conceptualization, J.T.; methodology, J.T. and A.C.; formal analysis, J.T., A.C. and D.F.; investigation, J.T. and A.C.; writing—original draft preparation, J.T, A.C. and D.F.; writing—review and editing, J.T., A.C. and D.F. All authors have read and agreed to the published version of the manuscript.

Funding: This research received no external funding.

Acknowledgments: None.

Conflicts of Interest: The authors declare no conflicts of interest.

References

1. Dehghani, A.; Aslani, F. A review on defects in steel offshore structures and developed strengthening techniques. *Structures* **2019**, *20*, 635–657. [CrossRef]
2. Price, S.J.; Figueira, R.B. Corrosion protection systems and fatigue corrosion in offshore wind structures: Current status and future perspectives. *Coatings* **2017**, *7*, 25. [CrossRef]
3. Papatheocharis, T.; Saravanis, G.C.; Perdikaris, P.C.; Karamanos, S.A. Fatigue of welded tubular X-joints in offshore wind platforms. In *International Conference on Offshore Mechanics and Arctic Engineering, Proceedings of the ASME 2019 38th International Conference on Ocean, Offshore and Arctic Engineering, Glasgow, Scotland, UK, 9–14 June 2019*; Materials Technology; ASME: New York, NY, USA, 2019; Volume 4. [CrossRef]
4. Kong, X.; Lv, J.; Gao, N.; Peng, X.; Zhang, J. An experimental study of galvanic corrosion on an underwater weld joint. *J. Coast. Res.* **2018**, *84*, 63–68. [CrossRef]
5. Liang, L.; Pang, Y.; Zhu, Z.; Tang, Y.; Xiang, Y. Influencing factors of various combinations of abrasion, cavitation and corrosion caused by multiphase flow impact. *Trans. Can. Soc. Mech. Eng.* **2019**, *43*, 130. [CrossRef]
6. Momber, A.W.; Marquardt, T. Protective coatings for offshore wind energy devices (OWEAs): A review. *J. Coat. Technol. Res.* **2018**, *15*, 13–40. [CrossRef]
7. López-Ortega, A.; Arana, J.L.; Rodriguez, E.; Bayón, R. Corrosion, wear and tribocorrosion performance of a thermally sprayed aluminum coating modified by plasma electrolytic oxidation technique for offshore submerged components protection. *Corros. Sci.* **2018**, *143*, 258–280. [CrossRef]
8. Łatka, L.; Szala, M.; Michalak, M.; Pałka, T. Impact of atmospheric plasma spray parameters on cavitation erosion resistance of Al_2O_3-13% TiO_2 coatings. *Acta Phys. Pol. A* **2019**, *136*, 342–347. [CrossRef]
9. He, X.; Song, R.G.; Kong, D.J. Microstructure and corrosion behaviours of composite coatings on S355 offshore steel prepared by laser cladding combined with micro-arc oxidation. *Appl. Surf. Sci.* **2019**, *497*, 143703. [CrossRef]
10. Li, Y.; Li, C.; Tang, S.; Zheng, Q.; Wang, J.; Zhang, Z.; Wang, Z. Interfacial bonding and abrasive wear behavior of iron matrix composite reinforced by ceramic particles. *Materials* **2019**, *12*, 3646. [CrossRef]
11. Czupryński, A. Properties of Al_2O_3/TiO_2 and ZrO_2/CaO flame-sprayed coatings. *Mater. Tehnol./Mater. Technol.* **2017**, *51*, 205–212. [CrossRef]
12. Szala, M.; Dudek, A.; Maruszczyk, A.; Walczak, M.; Chmiel, J.; Kowal, M. Effect of atmospheric plasma sprayed TiO_2—10% NiAl cermet coating thickness on cavitation erosion, sliding and abrasive wear resistance. *Acta Phys. Pol. A* **2019**, *136*, 335–341. [CrossRef]
13. Czupryński, A.; Górka, J.; Adamiak, M.; Tomiczek, B. Testing of flame sprayed Al_2O_3 matrix coatings containing TiO_2. *Arch. Metall. Mater.* **2016**, *61*, 1363–1370. [CrossRef]
14. Adamiak, M.; Czupryński, A.; Kopyść, A.; Monica, Z.; Olender, M.; Gwiazda, A. The Properties of Arc-Sprayed Aluminum Coatings on Armor-Grade Steel. *Metals* **2018**, *8*, 142. [CrossRef]
15. Chmielewski, T.; Siwek, P.; Chmielewski, M.; Piątkowska, A.; Grabias, A.; Golański, D. Structure and selected properties of arc sprayed coatings containing in-situ fabricated Fe-Al intermetallic phases. *Metals* **2018**, *8*, 1059. [CrossRef]
16. Czupryński, A. Flame spraying of aluminum coatings reinforced with particles of carbonaceous materials as an alternative for laser cladding technologies. *Materials* **2019**, *12*, 3467. [CrossRef]

17. Górka, J.; Czupryński, A.; Żuk, M.; Adamiak, M.; Kopyść, A. Properties and structure of deposited nanocrystalline coatings in relation to selected construction materials resistant to abrasive wear. *Materials* **2018**, *11*, 1–15. [CrossRef]
18. Chmielewski, T.; Hudycz, M.; Krajewski, A.; Sałaciński, T.; Skowrońska, B.; Świercz, R. Structure investigation on titanium metallization coating deposited onto AlN ceramics substrate by means of friction surfacing proces. *Coatings* **2019**, *9*, 845. [CrossRef]
19. Liu, Y.; Li, C.X.; Huang, X.F.; Ma, K.; Luo, X.T.; Li, C.J. Effect of water environment on particle deposition of underwater cold spray. *Appl. Surf. Sci.* **2020**, *506*, 144542. [CrossRef]
20. Feng, X.; Cui, X.; Zheng, W.; Lu, B.; Dong, M.; Wen, X.; Zhao, Y.; Jin, G. Effect of the protective materials and water on the repairing quality of nickel aluminum bronze during underwater wet laser repairing. *Opt. Laser Technol.* **2019**, *114*, 140–145. [CrossRef]
21. Wen, X.; Jin, G.; Cui, X.; Feng, X.; Lu, B.; Cai, Z.; Zhao, Y.; Fang, Y. Underwater wet laser cladding on 316L stainless steel: A protective material assisted method. *Opt. Laser Technol.* **2019**, *111*, 814–824. [CrossRef]
22. Fu, Y.; Guo, N.; Zhou, L.; Cheng, Q.; Feng, J. Underwater wire-feed laser deposition of the Ti–6Al–4V titanium alloy. *Mater. Des.* **2020**, *186*, 108284. [CrossRef]
23. Cevik, B. The effect of pure argon and mixed gases on microstructural and mechanical properties of S275 structural steel joined by flux-cored arc welding. *Kov. Mater.* **2020**, *56*, 81–87. [CrossRef]
24. Kik, T.; Moravec, J.; Nováková, I. Numerical simulations of X22CrMoV12-1 steel multilayer welding. *Arch. Metall. Mater.* **2019**, *64*, 1441–1448. [CrossRef]
25. Srisuwan, N.; Kumsri, N.; Yingsamphancharoen, T.; Kaewvilai, A. Hardfacing welded ASTM A572-based, high-strength low-alloy steel: Welding, characterization, and surface properties related to the wear resistance. *Metals* **2019**, *9*, 244. [CrossRef]
26. Triwanapong, S.; Angthong, A.; Kimapong, K. Interpass temperature affecting abrasive wear resistance of SMAW hard-faced weld metal on JIS-S50C carbon steel. *Mater. Sci. Forum* **2019**, *950*, 60–64. [CrossRef]
27. Wang, J.; Sun, Q.; Teng, J.; Feng, J. Bubble evolution in ultrasonic wave-assisted underwater wet FCAW. *Weld. J.* **2019**, *98*, 150–163. [CrossRef]
28. Guo, N.; Du, Y.; Maksimov, S.; Feng, J.; Yin, Z.; Krazhanovskyi, D.; Fu, Y. Study of metal transfer control in underwater wet FCAW using pulsed wire feed method. *Weld. World* **2018**, *62*, 87–94. [CrossRef]
29. Yang, Q.; Han, Y.; Jia, C.; Wu, J.; Dong, S.; Wu, C. Impeding effect of bubbles on metal transfer in underwater wet FCAW. *J. Manuf. Process.* **2019**, *45*, 682–689. [CrossRef]
30. Chen, H.; Guo, N.; Xu, K.; Xu, C.; Zhou, L.; Wang, G. In-situ observations of melt degassing and hydrogen removal enhanced by ultrasonics in underwater wet welding. *Mater.Des.* **2020**, 108482. [CrossRef]
31. Yan, C.; Kan, C.; Li, C.; Tian, S.; Bai, Y.; Xue, F. Experimental and numerical investigation on underwater wet welding of HSLA steel. *IOP Conf. Ser. Mater. Sci. Eng.* **2018**, *394*, 022016. [CrossRef]
32. Tomków, J.; Fydrych, D.; Rogalski, G.; Łabanowski, J. Effect of the welding environment and storage time of electrodes on the diffusible hydrogen content in deposited metal. *Rev. Metal.* **2019**, *55*, e140. [CrossRef]
33. Chen, H.; Gui, N.; Liu, C.; Zhang, X.; Xu, C.; Wang, G. Insight into hydrostatic pressure effects on diffusible hydrogen content in wet welding joints using in-situ X-ray imaging method. *Int. J. Hyd. Energy* **2020**, in press. [CrossRef]
34. Świerczyńska, A.; Fydrych, D.; Landowski, M.; Rogalski, G.; Łabanowski, J. Hydrogen embrittlement of X2CRNiMoCuN25-6-2- super duplex stainless steel welded joints under cathodic protection. *Constr. Build. Mater.* **2020**, *238*, 117697. [CrossRef]
35. Tomków, J.; Fydrych, D.; Rogalski, G.; Łabanowski, J. Temper bead welding of S460N steel in wet welding conditions. *Adv. Mater. Sci.* **2018**, *18*, 5–14. [CrossRef]
36. Gao, W.; Wang, D.; Cheng, F.; Di, X.; Den, C.; Xu, W. Microstructural and mechanical performance of underwater wet welded S355 steel. *J. Mater. Process. Technol.* **2016**, *238*, 333–340. [CrossRef]
37. ISO 2560-A. *Classification of Coated Rod Electrodes for Arc Welding of Unalloyed Steel and Fine-Grained Steel*; ISO: Geneva, Switzerland, 2010.
38. Tomków, J.; Fydrych, D.; Rogalski, G. Role of bead sequence in underwater welding. *Materials* **2019**, *12*, 3372. [CrossRef] [PubMed]
39. ASTM G 65-00. *Standard Test Method for Measuring Abrasion Using the Dry Sand/Rubber Wheel Apparatus*; West Conshohocken, PA, USA, 2016.

40. Szala, M.; Szafran, M.; Macek, W.; Marchenko, S.; Hejwowski, T. Abrasion resistance S235, C45, AISI 304 and Hardox 500 steels with usage of garnet, corundum and carborundum abrasives. *Adv. Sci. Technol. Res. J.* **2019**, *13*, 151–161. [CrossRef]
41. EN ISO 9015-1:2011. *Destructive Tests on Welds in Metallic Materials. Hardness Testing. Hardness Test on Arc Welded Joint*; ISO: Geneva, Switzerland, 2011.
42. EN ISO 17639:2013. *Destructive Tests on Welds in Metallic Materials. Macroscopic and Microscopic Examination of Welds*; ISO: Geneva, Switzerland, 2013.
43. Chand, N.; Neogi, S. Mechanism of material removal during three-body abrasion of FRF composite. *Tribol. Lett.* **1998**, *4*, 81. [CrossRef]
44. Singh, P.T.; Singla, A.K.; Singh, J.; Singh, K.; Gupta, M.K.; Ju, H.; Song, Q.; Liu, Z.; Pruncu, C.I. Abrasive wear behavior of cryogenically treated boron steel (30MnCrB4) used for rotavator blades. *Materials* **2020**, *13*, 436. [CrossRef]
45. Jankauskas, V.; Kreivaitis, R.; Milcius, D.; Baltusnikas, A. Analysis of abrasive wear performance of arc welded hard layers. *Wear* **2008**, *265*, 1626–1632. [CrossRef]
46. Sun, Y.L.; Obasi, G.; Hamelin, C.J.; Vasileiou, A.N.; Flint, T.F.; Balakrishnan, J.; Smith, M.C.; Francis, J.A. Effects of dilution on alloy content and microstructure in multi-pass steel welds. *J. Mater. Process. Technol.* **2019**, *265*, 71–86. [CrossRef]
47. Saida, K.; Bunda, K.; Ogiwara, H.; Nishimoto, K. Microcracking susceptibility in dissimilar multipass welds of Ni-base alloy 690 and low-alloy steel. *Weld. Int.* **2015**, *29*, 668–680. [CrossRef]
48. Coronado, J.J.; Caicedo, H.F.; Gómez, A.L. The effects of welding process on abrasive wear resistance for hardfacing deposits. *Tribol. Int.* **2009**, *42*, 745–749. [CrossRef]
49. EN-ISO 15614-1:2017. *Specification and Qualification of Welding Procedures for Metallic Materials—Welding Procedure Test—Part 1: Arc and Gas Welding of Steels and Arc Welding of Nickel and Nickel Alloys*; ISO: Geneva, Switzerland, 2017.
50. Guo, N.; Liu, D.; Guo, W.; Li, X.; Feng, J. Effect of Ni on microstructure and mechanical properties of underwater wet welding joint. *Mater. Des.* **2015**, *77*, 25–31. [CrossRef]
51. Cui, Z.; Bhattacharya, S.; Green, D.E.; Alpas, A.T. Mechanisms of die wear and wear-induced damage at the trimmed edge of high strength steel sheets. *Wear* **2019**, *426*, 1635–1645. [CrossRef]

© 2020 by the authors. Licensee MDPI, Basel, Switzerland. This article is an open access article distributed under the terms and conditions of the Creative Commons Attribution (CC BY) license (http://creativecommons.org/licenses/by/4.0/).

Article

Analysis of Iron Anchor Diseases Unearthed from Gudu Ruins in Xianyang City, Shaanxi Province, China

Bingjie Mai [1], Youlu Chen [2], Ying Zhang [1], Yongsheng Huang [2], Juanli Wang [1,*], Yuhu Li [1,*], Ming Cao [3,*] and Jing Cao [1,*]

1. Ministry of Education, Engineering Research Center of Historical and Cultural Heritage Protection, School of Materials Science and Engineering, Shaanxi Normal University, Xi'an 710119, China; maibingjie@snnu.edu.cn (B.M.); y26932022@163.com (Y.Z.)
2. Cultural Relics Restoration Department, Xianyang Museum, Xianyang 712000, China; skycyl666@163.com (Y.C.); huangyongsheng69@163.com (Y.H.)
3. Cultural Property Protection Center, Ningbo University of Finance & Economics, Ningbo 315175, China
* Correspondence: wangjuanli@snnu.edu.cn (J.W.); liyuhu@snnu.edu.cn (Y.L.); mcao1123@163.com (M.C.); jingcao@snnu.edu.cn (J.C.)

Abstract: Iron cultural relics are easily affected by environmental factors and can completely rust away. As early as the Qin Dynasty in ancient China, Xianyang Gudu was part of the most important transportation route to the West from ancient Chang'an; research into Xianyang Gudu has provided important information for understanding the historical changes in ancient China, East–West trade, and ancient boating technology. In this research, we use the iron anchors unearthed from the Gudu ruins in Xianyang City, Shaanxi Province, China as the research object; then, we used a scanning electron microscope–energy dispersive spectrometer (SEM-EDS), a high-resolution X-ray diffractometer (XRD), ion chromatography, and other methods to detect the corroded products of the iron anchors, and analyzed the iron anchor diseases in different preservation environments to explore the relationship between iron anchor disease and the preservation environment. This research found that the corroded products of the iron anchors contained the harmful tetragonal lepidocrocite (β-FeOOH) and that a high concentration of salt ions in the river channel accelerated the corrosion of the anchors; this analysis, based on the disease results, can provide a basis for the subsequent scientific restoration of iron anchors.

Keywords: iron anchor; corrosion product; iron relics; corrosion mechanism

1. Introduction

Cultural relics are precious due to their historical, artistic, scientific, and social value [1–3]. In 2002, when building Xianyang Lake, researchers discovered by accident the ruins of an old river embankment. After cleanup and excavation by archaeologists, stone embankments and a large number of cypress piles were found on the riverbed, as well as iron pillars, iron rings, iron anchors, and other equipment for anchoring boats inlaid on the steps. Through further research, the ruins here were found to be an ancient ferry from the Ming and Qing dynasties. The Gudu ruins are located on the north bank of the Weihe River (108.737242 E, 34.338084 N), Weicheng District, Xianyang City, Shaanxi Province, China; they are 230 m long from east to west and 7–10 m wide from north to south and occupy an area of about 2000 m². The Xianyang ancient ferry began in the Yin and Shang dynasties and prospered during the Qin and Han dynasties; it was called the "first ferry in the Qin Dynasty". According to documented records, when the pedestrians stood at the head of the ferry bridge, they could see the majestic and magnificent palace of the Qin dynasty to the west. For thousands of years, Xianyang Gudu has been part of the main artery of the Guanzhong area in ancient China from the west to the western regions and from the south to Bashu area, and was the first stop from the west out of Chang'an and the ancient Silk

Road. From the Qin dynasty to the Qing dynasty in ancient China, Gudu played a very important role in transportation.

The discovery of the Gudu ruins in Xianyang was of great significance in the research of ancient transportation, commerce, military, literature, and urban changes, and has attracted widespread attention from all walks of life. In total, 172 pieces of ironware were unearthed at the Gudu ruins in Xianyang, including 12 iron anchors, 10 of which were preserved in situ in the exhibition hall of the excavation site at the Gudu ruins. The early on-site survey investigation of the unearthed ironware revealed that the batch and age of those unearthed ironware are different, and that the corrosion on the surface is more complicated. The soil on the surface of the ironware was not thoroughly cleaned after excavation, resulting in a large amount of dry and hard soil attached to its surface; some ironware was severely rusted, and the peelings were flaky, scaly, layered, and powdery. In addition to surface rust, some ironware also had defects such as deformation, cracking, fracture, perforation, etc. and on some of the thinner parts of the iron structure there were structural fractures due to corrosion.

The comprehensive and scientific analysis and detection of iron cultural relics and their diseases by modern analysis and detection methods are the premise for the protection and restoration of ironware. In this research, multiple analysis methods [4–16] were used to detect the unearthed iron anchor decay products, surface condensation, and harmful salts, and the material of the iron anchor was analyzed. Based on the results of the disease investigation, an experimental basis is provided for subsequent corresponding protection schemes.

2. Materials and Methods

2.1. Sampling

The samples in this experiment were seriously rusted, and the scientific analysis and the detection of components and rusty decay products were able to identify the cause of iron anchor corrosion and the degree of rust, to provide a scientific basis for subsequent protection and restoration [17–19]. First, the digital image collection of cultural relics samples was conducted and the disease map was drawn. Then, the samples were collected and the iron material and rust products were analyzed. Finally, the types of cultural relics disease were comprehensively analyzed and scientific restoration methods were put forward (Figure 1).

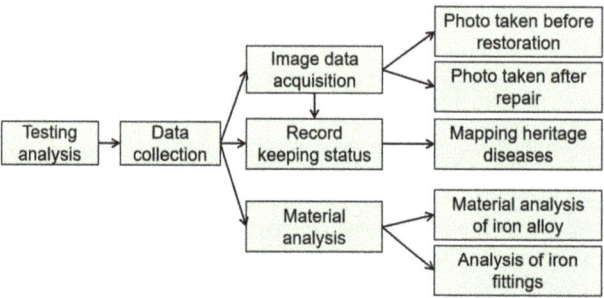

Figure 1. Flow chart describing the analysis and detection of iron relics.

2.2. Experimental Methods

Sample information: In order to ensure the maximum future and present aesthetic values of the iron anchor, this work analyzed the fallen fragments of samples from the anchor.

Ultra-depth of field optical microscopy: An ultra-depth of field optical microscope (KH-7700, Hirox Co., Ltd., Tokyo, Japan) was used to observe the macroscopic appearance of the sample; the magnification range was 20–120 times.

Metallographic analysis: the organization evaluation, process analysis, defect research and judgment of the raw materials of metal materials for processing structural parts and finished products.

SEM-EDS: SEM-EDS (SU3500, Hitachi High-Tech Co., Tokyo, Japan) was used to observe the microscopic morphology of the sample and detect the element composition and distribution. The sample was fixed on the platform with conductive adhesive, and the gold spraying time was set at 100 s to prepare the sample and eliminate the charge accumulation. The test conditions were vacuum mode, the acceleration voltage is 15 kV, and amplification was 100–300 times.

Ion chromatographic analysis: The ICS-90 ion chromatograph (Shanghai Leirui Scientific Instrument Co., Ltd, Shanghai, China) was used as the testing instrument. Grind the iron rust sample through an 80-mesh sieve, and dissolve 2 g in 100 mL of distilled water. Shake and soak for 7 days, and then detect the relevant ion content in the leachate solution. The measurement data analysis method adopted the general rules of JY/T020-1996 ion chromatography analysis method.

Laser Raman spectroscopy analysis: This analysis used the in Via Reflex micro-confocal laser Raman spectrometer (Renishaw, Gloucestershire, UK), with a scanning range of 100–4000 cm^{-1}; the signal-to-noise ratio of the third-order peak of detection silicon was greater than 22:1; spectral resolution for the full visible spectrum was ≤ 1 cm^{-1}.

High-resolution X-ray diffractometer: X'Pert Pro MPD X-ray diffractometer (Smart Lab (9), Rigaku Co., Tokyo, Japan) was used as the analytical instrument; the analysis condition adopts the anode target as the copper target; the test voltage is 40 KV; the test current is 30 mA; the 2θ range is $5°–90°$; step width, $0.02°$; scanning rate, $4°/min$. The experimental data analysis method was based on "JY/T009-1996 General Principles of Rotating Target Polycrystalline X-ray Diffraction Method"; the X-ray diffraction experiment was carried out using a powder sample method, and the diffraction spectrum was analyzed by Jade 7.0.

Silver nitrate titration detection of rusty samples: Take the rusty products and grind them with agate to powder. Dissolve the sample with nitric acid. After 2 min, filter with filter paper to obtain a clear solution, then add 2–3 drops of $AgNO_3$ solution to the clear solution and observe whether white flocs appear. The nitric acid solution used was a dilute nitric acid solution prepared in a ratio of 2:1 to distilled water and nitric acid solution. The concentration of $AgNO_3$ was 1%. Observe the above reactions.

3. Results and Discussion

3.1. Investigation and Survey of the Occurrence Environment

Xianyang City, Shaanxi Province, China is located in a warm temperate zone with a continental monsoon climate; the four seasons are clearly cold, hot, dry, and wet; the climate is mild; light, heat, and water resources are abundant; the annual average precipitation is 537–650 mm; and the average temperature is 9.0–13.2 °C. The topography of the Gudu ruins area is higher in the north and lower in the south, with a ground elevation of 378.0–397.5 m; the landform is a river valley accumulation landform. The south of the Gudu ruins relies on Xianyang Lake, the normal water level of the lake is 382.5 m, and the elevation of the Gudu ruins pit is lower than the water level of Xianyang Lake. The humidity in the ruins is above 70%. The Gudu site is a specific place formed by wading through the river with boats in early civilization to transport people and objects to the opposite bank (Figure 2). After arriving at the destination, the anchor can fix and stabilize the position of the boat. During the use of the iron anchor in the riverway water environment, the iron anchor was corroded by various ions, and the desalination protection treatment was not carried out in time after the excavation; thus, some of the iron anchors left at the excavation site contacted with the sand and salt. At the same time, the ruins relied on Xianyang Lake, and the water level of the lake changed, which caused some seepage in the ruins pit. The various chemical components contained in the seepage water made the disease more serious. Another part of the ironware is placed in the environment of the exhibition hall, and the temperature and humidity of the exhibition hall are greatly affected by weather changes; the ironware is mixed with other types of cultural relics, and the preservation environment is relatively harsh.

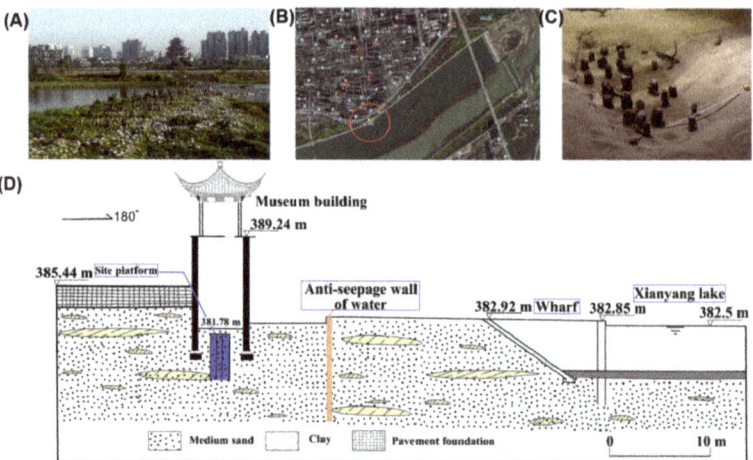

Figure 2. Occurrence conditions of Gudu ruins. (**A**): excavated photos of the Gudu ruins; (**B**): geomorphology around the Gudu ruins; (**C**): site photos of the Gudu ruins; (**D**): geological section of Gudu ruins area.

3.2. Preservation Status and Disease Type

The unearthed anchor ironware is stored in the exhibition hall of the Gudu ruins excavation site; the rest are displayed in the cultural relics exhibition hall of Xianyang Museum. The preservation conditions in the exhibition hall are simple and the ventilation equipment cannot be used for 24 h, which causes the exhibition hall humidity and temperature during the rainy season to be maintained at a high level. The on-site survey and investigation found that 91% of the ironware in this batch had different degrees of rust; 56% had severe rust, 21% had cracks and deformations, and 12% had severe damage. In total, 12 iron anchors were unearthed from the Gudu ruins and 10 iron anchors were preserved in situ in the ruins pit and the remaining iron anchors are located in the exhibition hall of the Gudu Museum. Due to various factors in the riverway environment, the iron anchors were corroded by chloride ions, and severe corrosion occurred; after unearthing, the desalination protection treatment was not carried out in time, and the chlorine in the ironware and the water and oxygen in the air further reacted with the iron, accelerating the corrosion of the ironware. At the same time, we drew the disease map of this batch of anchors. Four randomly selected anchors showed similar defects, including corrosion, iron tumor, soil attachments, welding marks, and layer stripping, as shown in Figure 3A–D.

3.3. Cast Iron Material Composition of Iron Anchor

Through metallographic examination, scanning electron microscopy (Figure 4A,B), and energy spectrum analysis (Figure 4C), the metallographic structure of the iron anchor corrosion sample was determined as ferrite with a small amount of massive cementite. This illustrates that the cast iron material of this batch of iron anchors is white iron.

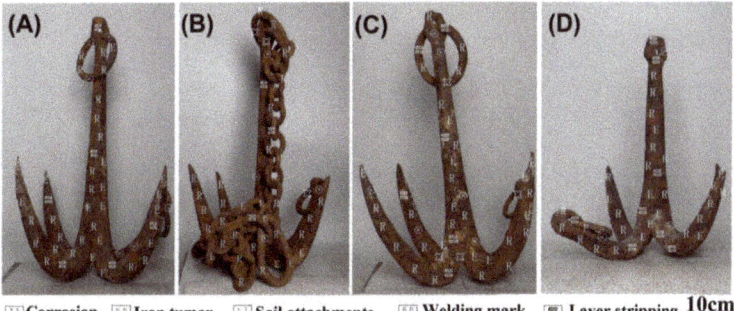

Figure 3. Typical diseases of the anchors: (**A**–**D**) are disease maps of four samples.

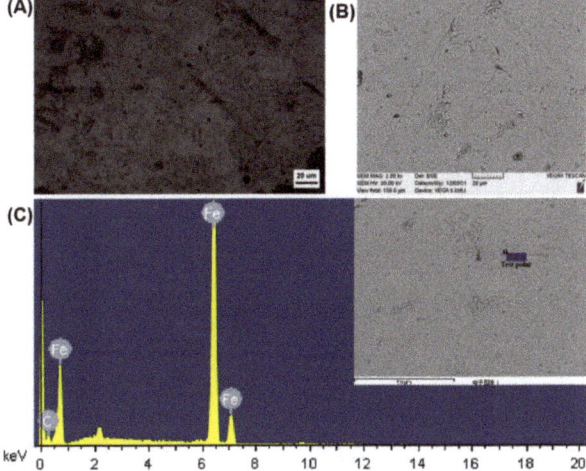

Figure 4. Metallographic analysis of iron anchor samples. (**A**) Metallographic microscopy characterizes the properties of metals; (**B**) scanning metallography of anchor; (**C**) EDX energy spectrum of anchor.

3.4. Corroded Microstructure

As shown in the Figure 5, there were some soil and loose rusty products on the surface of the iron anchor. A low density of rusty products on the surface of the iron anchor was observed, and an oxide film that can effectively protect the body was not formed on the surface of the iron (Figure 5A). There was rust and peeling on the surface of the iron anchor, and there are yellow rust and rusty spots, which are presumed to be the surface corroded products caused by the presence of soluble chloride salts in the sample (Figure 5B). As shown in Figure 5C, the appearance of the sample surface corrosion products is flaky, and the surface corrosion products were tan, presumably because the iron surface chloride soluble salt reacted with the iron surface generated by the brown rust products; in the subsequent experiments on the samples of chloride corrosion products and titration analysis, types of chloride and chloride content were identified. There are soil adhesion and yellow–brown rusty products on the surface of the sample iron anchor in Figure 5D, from which it can be inferred that it may be because the unearthed area of the sample is an ancient riverbed area, and some river sand residues are attached to its surface. Due to the abundant groundwater around the excavated area, the leaching effect of groundwater on the surrounding rock formations results in a high content of soluble chloride salts in the soil around the riverbed, leading to the formation of chemical corrosion on the surface iron anchors.

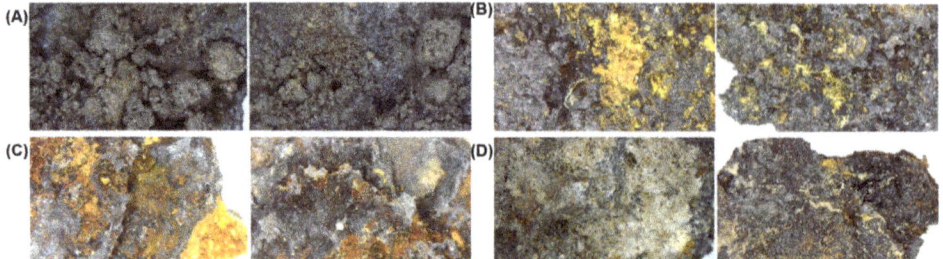

Figure 5. Iron anchor photomicrograph. (**A–D**) are four sample micrographs.

The scanning electron micrographs show that the microscopic morphology of the rusty products in the samples is different; the corrosion products on the surface of the samples are lumpy (Figure 6A), granular (Figure 6B), needle-like (Figure 6C), and spherical (Figure 6D), all of which have loose surface structures. This may be due to the different elemental composition of the sample iron and the different surface storage environment. The oxide layer formed on the surface has a low density, and the existence of a large number of holes means that it cannot effectively cover the inner layer, nor can it prevent further corrosion caused by the infiltration of O_2, H_2O and Cl^-, which leads to the deterioration of the surface of the iron anchor.

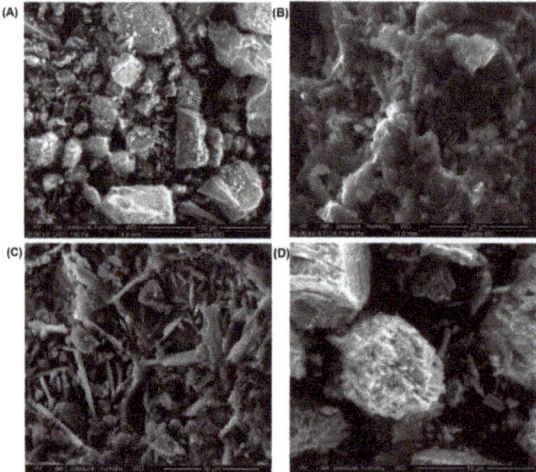

Figure 6. Electron microscope image of corrosion products of iron anchor block. (**A–D**) are four sample micrographs. (**A**) Bulk corrosion products; (**B**) granular corrosion products; (**C**) needle-like corrosion products; (**D**) spherical corrosion products.

3.5. Corrosion Products

The EDX energy spectrum analysis showed that the main components were silicon dioxide (Figure 7A), ferric oxide (Figure 7B), ferric oxide (Figure 7C), ferrous oxide, and calcium and magnesium carbonate (Figure 7D). Under the buried environment of river channels, silicon dioxide, calcium and magnesium carbonate appeared on iron anchors. Elemental iron is not very reductive, but in the presence of Fe, it is rapidly oxidized from a zero-valent state to a high-valence state. The corrosion of iron is a continuous process. First, iron loses two electrons to become Fe^{2+}, and the main reactions are hydrogen evolution corrosion or oxygen absorption corrosion. We found that the soil pH ranged from 6.13 to 6.38 from sampling soil sand and gravel from Gudu ruins. When the iron is immersed in water or there is a layer of acidic water film on the surface, Fe is more active than H^+, and the substitution reaction will occur to form Fe^{2+} and H_2. The main Fe^{2+} compounds were ferrous oxide and ferrous hydroxide, which continue to react with oxidizing substances to form Fe_3O_4, Fe_2O_3, and iron hydroxide [20–22]. Consequently, we detected Fe_2O_3, FeO_3 and FeO on the corroded surface of the anchor.

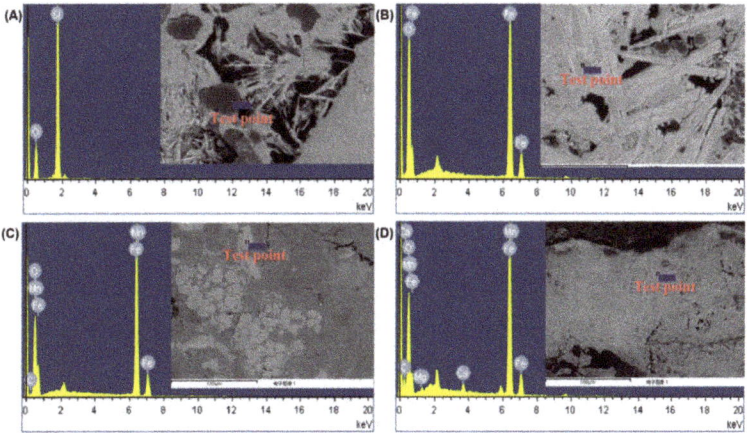

Figure 7. EDX spectrum. The iron anchor samples detected silica (**A**), ferric oxide (**B**), ferric oxide (**C**), and ferrous oxide and calcium–magnesium carbonate (**D**).

In Figure 8A, the actual peak positions on the diffraction pattern were: 21.2°, 26.6° and 36.7°, corresponding to the α-FeOOH and γ-FeOOH peak positions in the standard library, which indicated that the sample contains goethite (α-FeOOH) and lepidocrocite (γ-FeOOH); 27.1° and 50.4°, corresponding to the silica peak positions in the standard library, which indicated that the sample contains silica; 39.9° and 54.2°, corresponding to the peak positions of γ-Fe_2O_3 in the standard library, which indicated that the sample contains γ-Fe_2O_3; 43.1°, corresponding to the peak positions of Fe_3O_4 in the standard library, which indicated that the sample contains Fe_3O_4. As shown Figure 8B, the peak locations at 21.2°, 26.6°, and 36.7° indicated that the sample contains α-FeOOH and γ-FeOOH; 50.4°, corresponding to the silica peak positions in the standard library, which indicated that the sample contains silica; 39.9° and 54.2°, corresponding to the peak positions of γ-Fe_2O_3 in the standard library, which indicated that the sample contains γ-Fe_2O_3. The peak locations at 21.2° and 26.6° in Figure 8C indicated that the sample contains α-FeOOH and γ-FeOOH. Figure 8C showed a characteristic peak in 44.7° of silica. In Figure 8D, the actual peak positions on the diffraction pattern were 26.6° and 43.1°, corresponding to the α-FeOOH peak positions in the standard library, which indicated that the sample contains goethite (α-FeOOH). Figure 8D also showed a characteristic peak of silica, similar to Figure 8C. From the above four typical samples, it can be seen that the rusty product samples of the sampled iron anchors contain goethite (α-FeOOH), hematite (α-Fe_2O_3), maghemite (γ-Fe_2O_3), akaganeite (β-FeOOH), lepidocrocite (γ-FeOOH), and magnetite (Fe_3O_4), as well as a small amount of hexagonal lepidocrocite. The lepidocrocite (γ-FeOOH) is unstable and, under certain conditions, can be transformed into goethite (α-FeOOH) and magnetite (Fe_3O_4). This showed that the iron anchor corroded product samples were all the corroded products with different combinations of corrosion components and a serious degree of corrosion. Silica (SiO_2) was also detected in the sample; this substance is the main component of soil and sand, and the presemnce of this substance may be due to the excavation location of the batch of ironware in the sand riverway, and a large amount of sand adhered to the surface of the unearthed ironware to form the surface pollutants and hardened materials. When cleaning, sandblasting technology and sanding machines can be used to physically remove silica.

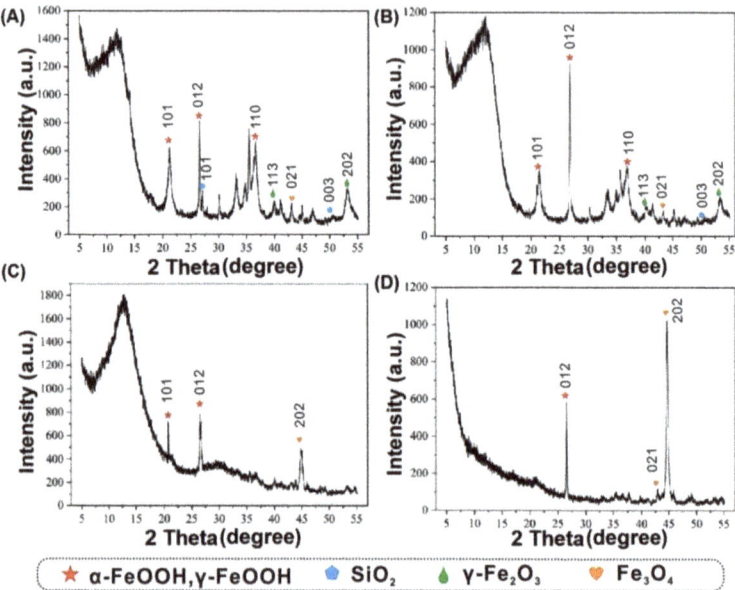

Figure 8. XRD spectrum. (A–D) XRD maps of four sample.

It can be seen from the Figure 9 that the peak at 428 cm^{-1} of samples in Figure 9A,B was consistent with the 400 cm^{-1} peak of α-FeOOH. The peaks at 226 and 278 cm^{-1} were consistent with those at 230 and 250 cm^{-1} of α-Fe$_2$O$_3$. Therefore, it is speculated that goethite (α-FeOOH) and hematite (α-Fe$_2$O$_3$) may be present in the sample. The peaks at 221 and 257 cm^{-1} in Figure 9C,D were consistent with those at 230 and 250 cm^{-1} in α-Fe$_2$O$_3$. The peak at 240 cm^{-1} was consistent with that of γ-FeOOH at 245cm^{-1}, so it was speculated that the sample may contain hematite (α-Fe$_2$O$_3$) and fibroite (γ-FeOOH). The Raman spectrum showed that the corroded products mainly contain goethite (α-FeOOH), lepidocrocite (γ-FeOOH), and hematite (α-Fe$_2$O$_3$), and most of the samples contain multiple substances, with clear characteristic peaks. It is shown that these samples are rusty products with different combinations of rusty components and serious degrees of rust.

3.6. Chloride Ions in the Sample

After the four samples were titrated with silver nitrate solution (Figure 10), white flocculent precipitates appeared, indicating that there were different concentrations of chloride and soluble chloride salts in the four samples, and the content of white precipitate in sample No. 2 is greater than that in other samples, confirming that sample No. 2 had a higher chloride content, as shown by the ion chromatography analysis.

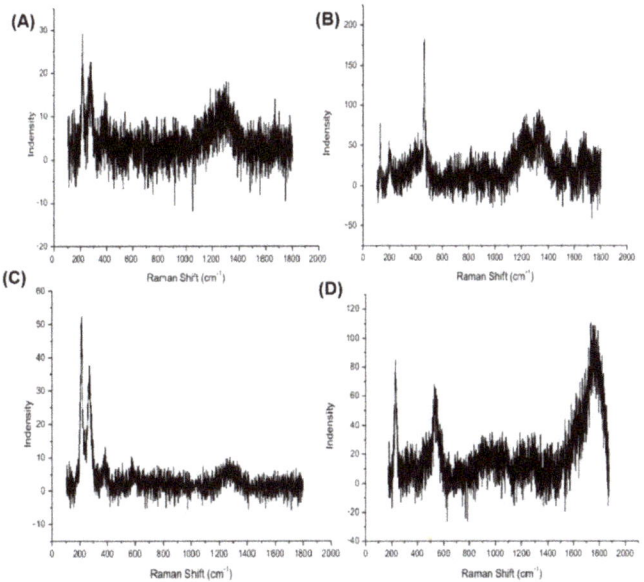

Figure 9. Raman spectrogram. (**A–D**) Raman maps of four sample.

Sample	Cl⁻	NO₃⁻	SO₄²⁻
1	0.018%	0.033%	0.18%
2	0.056%	0.052%	0.27%
3	0.030%	0.027%	0.20%
4	0.037%	0.027%	0.60%

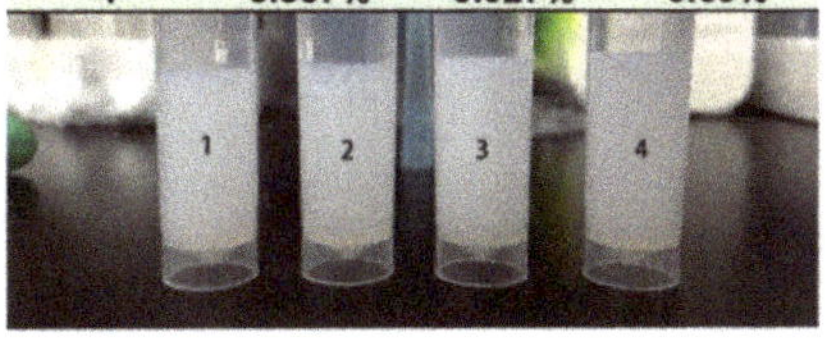

Figure 10. Ion chromatography and titrating with silver nitrate solution results.

3.7. Mechanism of Corrosion Evolution

There are many reasons for iron corrosion, including uneven structure, more impurities, and the unstable properties of iron itself. In addition, the protection of iron relics is also greatly influenced and restricted by the environment [23,24]. In the corrosion process of ironware cultural relics, the chemical corrosion, electrochemical corrosion and microbial corrosion often occur at the same time on the surface of ironware [25,26].

Iron anchors unearthed at the Gudu ruins were in a state of riverway water corrosion; the chloride ions in the river water have a destructive effect on the surface oxide film and accelerate the dissolution of the surface film [27–29]. After finally sinking into the sand, the Gudu ruins were immersed in an environment rich in groundwater for a long time, causing the iron anchor to continue to react with the chloride ions in the water, which promoted the dissolution of the film surface and was destroyed.

As mentioned earlier, the soil environment of Gudu ruins was weakly acidic; when the water film in contact with the iron surface becomes acidic, the active order of iron is ahead of H^+, and substitution reactions occur to form Fe^{2+} and H_2 [30,31]. While, Gudu ruin was located below the level of Xianyang Lake in all seasons, presenting a humid environment. In the humid environment of the buried environment, the iron cultural relics on the basis of chemical corrosion and electrochemical corrosion would also participate.

Metallurgical microscopy (Figure 4) showed the presence of ferrite and cementite on the corroded surface. When the oxygen-containing water film is formed on the surface of the iron anchor, there are a certain amount of electrolyte ions (Cl$^-$, Na$^+$), and the ferrite (α-Fe) and cementite (Fe$_3$C) will form numerous microgalvanic cells, thus forming microcell corrosion. As shown in Figure 11, the electrolyte water film on the surface of the iron anchor is the external circuit, Fe is the internal anode, and Fe$_3$C is the cathode. Compared with the single chemical corrosion, the electrochemical reactions could greatly accelerate the corrosion of iron.

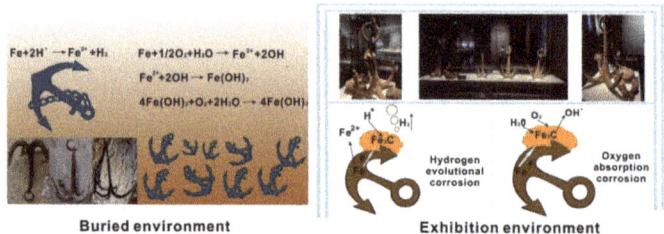

Figure 11. Corrosion mechanism of iron anchor in different environments.

Moreover, in the corrosive medium of the atmosphere, soil and water with a high salt content, especially a high chloride content, will destroy the stable α-Fe$_2$O$_3$ oxide film on the surface of ironware, which will accelerate the corrosion of iron. In the buried environment, the main chloric corrosion product of iron cultural relics is FeCl$_3$, which is concentrated in the form of an acidic solution between iron matrix and rust layer, as well as in the holes and crevices of the rust layer. When the iron relics were unearthed, FeCl$_3$ was oxidized and hydrolyzed in the air, producing FeOOH: 4FeCl + O$_2$ + 6H$_2$O → 4FeOOOH + 8HCl. The FeOOH produced by the reaction may be α-FeOOH, β-FeOOH, or γ-FeooH. This is because the chloride can accelerate local corrosion, such as pitting, stress corrosion, inter-granular corrosion, and crevice corrosion, and Cl$^-$ could prevent the conversion of active γ-FeOOH generated on the surface of steel to inactive α-FeOOH [32–34]. However, the corrosion of these anchors was not explored and further protected over time, which resulted in salt ions exacerbating the corrosion of anchors after they were unearthed.

4. Conclusions

The iron cultural relics are a type of cultural relics that are difficult to preserve in the existing cultural relics. Due to their special physical and chemical properties, they are relatively small in number and are easily rusted due to environmental factors. This research took the iron anchors unearthed from the Gudu ruins in Xianyang City, Shaanxi Province, China as the research object. The disease and corrosion mechanisms of the iron anchors unearthed were researched by XRD, EDX, Raman spectroscopy, ion chromatography, and nitrate titration, to explore the storage environment and the corrosion before and after the excavation and the change of the relationship between the storage environment and the anchor disease. The research found that iron anchors unearthed at the Gudu ruins were in a state of riverway water corrosion, and the abundant chloride ions in the river intensified the corrosion of iron anchors before they were excavated, and the Gudu ruin was located below the level of Xianyang Lake in all seasons, presenting a humid environment, which further promoted the electrochemical corrosion of iron anchors after they were excavated. However, the corrosion of these anchors was not explored or further protected over time, which resulted in salt ions aggravating the corrosion of anchors after they were unearthed, and the electrochemical reaction could greatly accelerate the corrosion of iron in the humid exhibition environment. Therefore, the subsequent protection and repair of these anchors should be derusting, desalting, and cleaning as soon as possible, to prevent acceleration of the destruction of cultural relics, and the research on the correlation between environment and deterioration of iron relics is very important to scientifically formulate a route for protection technology.

Author Contributions: Conceptualization, B.M., J.W., M.C. and J.C.; Data curation, Y.C., Y.H. and M.C.; Formal analysis, M.C.; Funding acquisition, B.M., J.W. and J.C.; Investigation, B.M., Y.C., Y.Z. and Y.H.; Methodology, J.W., Y.L. and M.C.; Project administration, Y.L. and J.C.; Software, Y.Z.; Writing—original draft, B.M.; Writing—review & editing, J.W. and J.C. All authors have read and agreed to the published version of the manuscript.

Funding: The authors gratefully acknowledge the financial support by the National Natural Science Foundation of China (No. 22102094), the Fundamental Research Funds for the Central Universities (GK 202103061 and GK 202103058), and the Key Research and Development Program of Shaanxi Province, China (No. 2021SF-457).

Institutional Review Board Statement: Not applicable.

Informed Consent Statement: Not applicable.

Data Availability Statement: The datasets used and/or analysis results obtained in the current study are available from the corresponding author on request.

Acknowledgments: The authors thank Xianyang Museum and Xianyang Gudu Ruin Museum of China for their support and help in this project.

Conflicts of Interest: The authors declare that they have no conflict of interest related to this work. We declare that we do not have any commercial or associated interest that represents a conflict of interest in connection with the work submitted.

References

1. Song, L.; Li, X.; Yang, Y.G.; Zhu, X.; Guo, Q.; Liu, H. Structured-light based 3D reconstruction system for cultural relic packaging. *Sensors* **2018**, *18*, 2981. [CrossRef] [PubMed]
2. Li, J.; Zhang, X.; Xiao, L.; Liu, K.; Li, Y.; Zhang, Z.; Chen, Q.; Ao, X.; Liao, D.; Gu, Y.; et al. Changes in soil microbial communities at Jinsha earthen site are associated with earthen site deterioration. *BMC Microbiol.* **2020**, *20*, 147. [CrossRef] [PubMed]
3. Sun, M.; Zhang, F.; Huang, X.; Han, Y.; Jiang, N.; Cui, B.; Guo, Q.; Kong, M.; Song, L.; Pan, J. Analysis of microbial community in the archaeological ruins of Liangzhu city and study on protective materials. *Front. Microbiol.* **2020**, *11*, 684. [CrossRef]
4. Zhou, W.; Gan, Q.; Ji, J.; Yao, N.; Wang, J.; Zhou, Z.; Qi, X.; Shi, J. Non-destructive identification of pigments printed on six imperial China engraved coiling dragon stamps. *J. Raman Spectrosc.* **2016**, *47*, 316–320. [CrossRef]
5. Wang, X.; Zhen, G.; Hao, X.; Tong, T.; Ni, F.; Wang, Z.; Jia, J.; Li, L.; Tong, H. Spectroscopic investigation and comprehensive analysis of the polychrome clay sculpture of hua yan temple of the liao dynasty. *Spectrochim. Acta Part A Mol. Biomol. Spectrosc.* **2020**, *240*, 118574. [CrossRef] [PubMed]
6. Hunt, A.; Thomas, P.; James, D.; David, B.; Geneste, J.-M.; Delannoy, J.-J.; Stuart, B. The characterisation of pigments used in X-ray rock art at dalakngalarr 1, central-western arnhem land. *Microchem. J.* **2016**, *126*, 524–529. [CrossRef]
7. Ilmi, M.M.; Nurdini, N.; Maryanti, E.; Saiyasombat, C.; Setiawan, P.; Kadja, G.T.M. Ismunandar multi-analytical characterizations of prehistoric rock art pigments from karim cave, sangkulirang–mangkalihat site, east kalimantan, indonesia. *Microchem. J.* **2020**, *155*, 104738. [CrossRef]
8. Zaffino, C.; Guglielmi, V.; Faraone, S.; Vinaccia, A.; Bruni, S. Exploiting external reflection FTIR spectroscopy for the in-situ identification of pigments and binders in illuminated manuscripts. Brochantite and posnjakite as a case study. *Spectrochim. Acta Part A Mol. Biomol. Spectrosc.* **2015**, *136 Pt B*, 1076–1085. [CrossRef]
9. Fu, P.; Teri, G.; Li, J.; Huo, Y.; Yang, H.; Li, Y. Analysis of an ancient architectural painting from the jiangxue palace in the imperial museum, Beijing, China. *Anal. Lett.* **2020**, *54*, 684–697. [CrossRef]
10. Moyo, S.; Mphuthi, D.; Cukrowska, E.; Henshilwood, C.S.; van Niekerk, K.; Chimuka, L. Blombos cave: Middle stone age ochre differentiation through FTIR, ICP OES, ED XRF and XRD. *Quat. Int.* **2016**, *404*, 20–29. [CrossRef]
11. Li, J.; Mai, B.; Fu, P.; Teri, G.; Li, Y.; Cao, J.; Li, Y.; Wang, J. Multi-analytical research on the caisson painting of Dayu temple in Hancheng, Shaanxi, China. *Coatings* **2021**, *11*, 1372. [CrossRef]
12. Rosi, F.; Burnstock, A.; Berg, K.; Miliani, C.; Brunetti, B.G.; Sgamellotti, A. A non-invasive XRF study supported by multivariate statistical analysis and reflectance FTIR to assess the composition of modern painting materials. *Spectrochim. Acta Part A Mol. Biomol. Spectrosc.* **2009**, *71*, 1655–1662. [CrossRef] [PubMed]
13. Bisegna, F.; Ambrosini, D.; Paoletti, D.; Sfarra, S.; Gugliermetti, F. A qualitative method for combining thermal imprints to emerging weak points of ancient wall structures by passive infrared thermography—A case study. *J. Cult. Herit.* **2014**, *15*, 199–202. [CrossRef]

14. Kubik, M. Chapter 5 hyperspectral imaging: A new technique for the non-invasive study of artworks. In *Physical Techniques in the Study of Art, Archaeology and Cultural Heritage*; Elsevier: Amsterdam, The Netherlands, 2007; Volume 2, pp. 199–259. [CrossRef]
15. Ricci, M.; Laureti, S.; Malekmohammadi, H.; Sfarra, S.; Lanteri, L.; Colantonio, C.; Calabrò, G.; Pelosi, C. Surface and interface investigation of a 15th century wall painting using multispectral imaging and pulse-compression infrared thermography. *Coatings* **2021**, *11*, 546. [CrossRef]
16. Elias, M.; Mas, N.; Cotte, P. Review of several optical non-destructive analyses of an easel painting. Complementarity and crosschecking of the results. *J. Cult. Herit.* **2011**, *12*, 335–345. [CrossRef]
17. Tao, N.; Wang, C.; Zhang, C.; Sun, J. Quantitative measurement of cast metal relics by pulsed thermal imaging. *Quant. Infrared Thermogr. J.* **2020**, *5*, 27–40. [CrossRef]
18. Ambrosini, D.; Paoletti, A.; Paoletti, D.; Sfarra, S. NDT methods in artwork corrosion monitoring. In *O3A: Optics for Arts, Architecture, and Archaeology*; SPIE: Philadelphia, PA, USA, 2007; pp. 327–335. [CrossRef]
19. Xiaoli, L.; Ning, T.; Sun, J.G.; Yong, L.; Liang, Q.; Fei, G.; Yi, H.; Guang, W.; Lichun, F. Evaluation of an ancient cast-iron Buddha head by step-heating infrared thermography. *Infrared Phys. Technol.* **2019**, *98*, 223–229. [CrossRef]
20. Motte, R.D.; Basilico, E.; Mingant, R.; Kittel, J.; Ropital, F.; Combrade, P.; Necib, S.; Deydier, V.; Crusset, D.; Marcelin, S. A study by electrochemical impedance spectroscopy and surface analysis of corrosion product layers formed during CO_2 corrosion of low alloy steel. *Corros. Sci.* **2020**, *172*, 108666. [CrossRef]
21. Remazeilles, C.; Neff, D.; Bourdoiseau, J.A.; Sabot, R.; Jeannin, M.; Refait, P. Role of previously formed corrosion product layers on sulfide-assisted corrosion of iron archaeological artefacts in soil. *Corros. Sci.* **2017**, *129*, 169–178. [CrossRef]
22. Samide, A.; Tutunaru, B.; Dobritescu, A.; Negrila, C. Study of the corrosion products formed on carbon steel surface in hydrochloric acid solution. *J. Therm. Anal.* **2012**, *110*, 145–152. [CrossRef]
23. Tamura, H. The role of rusts in corrosion and corrosion protection of iron and steel. *Corros. Sci.* **2008**, *50*, 1872–1883. [CrossRef]
24. Novakova, A.A.; Gendler, T.S.; Manyurova, N.D.; Turishcheva, R.A. A mössbauer spectroscopy study of the corrosion products formed at an iron surface in soil. *Corros. Sci.* **1997**, *39*, 1585–1594. [CrossRef]
25. Watkinson, D.E.; Rimmer, M.B.; Emmerson, N.J. The influence of relative humidity and intrinsic chloride on post-excavation corrosion rates of archaeological wrought iron. *Stud. Conserv.* **2019**, *64*, 456–471. [CrossRef]
26. Carlin, W.; Keith, D.; Rodriguez, J. Less is more: Measure of chloride removal rate from wrought iron artifacts during electrolysis. *Stud. Conserv.* **2001**, *46*, 68–76. [CrossRef]
27. Song, Y.; Jiang, G.; Chen, Y.; Zhao, P.; Tian, Y. Effects of chloride ions on corrosion of ductile iron and carbon steel in soil environments. *Sci. Rep.* **2017**, *7*, 6865. [CrossRef]
28. Liu, T.M.; Wu, Y.H.; Luo, S.X.; Sun, C. Effect of soil compositions on the electrochemical corrosion behavior of carbon steel in simulated soil solution. *Mater. Werkst.* **2010**, *41*, 228–233. [CrossRef]
29. Chen, J.; Chen, Z.; Ai, Y.; Xiao, J.; Pan, D.; Li, W.; Huang, Z.; Wang, Y. Impact of soil composition and electrochemistry on corrosion of rock-cut slope nets along railway Lines in China. *Sci. Rep.* **2015**, *5*, 14939. [CrossRef]
30. Viereck, S.; Jovanovic, Z.R.; Haselbacher, A.; Steinfeld, A. Investigation of Na_2SO_4 removal from a supercritical aqueous solution in a dip-tube salt separator. *J. Supercrit. Fluids* **2017**, *133*, 146–155. [CrossRef]
31. Galvan-Reyes, C.; Fuentes-Aceituno, J.C.; Salinas-Rodríguez, A. The role of alkalizing agent on the manganese phosphating of a high strength steel part 1: The individual effect of naoh and NH_4OH. *Surf. Coat. Technol.* **2016**, *291*, 179–188. [CrossRef]
32. Wang, Y.; Kong, G.; Che, C.; Zhang, B. Inhibitive effect of sodium molybdate on the corrosion behavior of galvanized steel in simulated concrete pore solution. *Constr. Build. Mater.* **2018**, *162*, 383–392. [CrossRef]
33. Ge, C.Y.; Yang, X.G.; Hou, B.R. Synthesis of polyaniline nanofiber and anticorrosion property of polyaniline–epoxy composite coating for q235 steel. *J. Coat. Technol. Res.* **2012**, *9*, 59–69. [CrossRef]
34. Ye, Y.; Liu, Z.; Liu, W.; Zhang, D.; Zhao, H.; Wang, L.; Li, X. Superhydrophobic oligoaniline-containing electroactive silica coating as pre-process coating for corrosion protection of carbon steel. *Chem. Eng. J.* **2018**, *348*, 940–951. [CrossRef]

Review

Plasma Electrolytic Oxidation Ceramic Coatings on Zirconium (Zr) and ZrAlloys: Part I—Growth Mechanisms, Microstructure, and Chemical Composition

Navid Attarzadeh [1,2] and Chintalapalle V. Ramana [1,3,*]

1. Center for Advanced Materials Research, University of Texas at El Paso, 500 W. Univ. Ave., El Paso, TX 79968, USA; nattarzadeh@utep.edu
2. Environmental Science and Engineering, University of Texas at El Paso, 500 W. Univ. Ave., El Paso, TX 79968, USA
3. Department of Mechanical Engineering, University of Texas at El Paso, 500 W. Univ. Ave., El Paso, TX 79968, USA
* Correspondence: rvchintalapalle@utep.edu

Abstract: Recently, a significant number of research projects have been directed towards designing and developing ceramic coatings for zirconium-based substrates due to their outstanding surface properties and utilization in modern technologies. The plasma electrolytic oxidation (PEO) coating is an environmentally friendly wet coating method that can be performed in a wide range of electrolytes. The surface characteristics of PEO coatings can be tailored by changing electrochemical parameters, electrolyte composition, and substrate alloying elements to adopt a conformal and adhesive PEO ceramic coating for the final demanding applications in chemical, electronics, and energy technologies. This review focuses on deriving a deeper fundamental understanding of the PEO growth mechanisms and the effect of process parameters on transient discharge behavior at breakdown, initiation, and growth of the oxide layer and incorporating species from the electrolyte. It highlights the fundamental microstructural properties associated with structural defects, phase transformation, and the role of additives.

Keywords: ceramic coatings; plasma electrolytic oxidation (PEO) coating; microstructure; growth mechanism; zirconium and zirconium-based alloys

Citation: Attarzadeh, N.; Ramana, C.V. Plasma Electrolytic Oxidation Ceramic Coatings on Zirconium (Zr) and ZrAlloys: Part I—Growth Mechanisms, Microstructure, and Chemical Composition. *Coatings* **2021**, *11*, 634. https://doi.org/10.3390/coatings11060634

Academic Editor: Małgorzata Norek

Received: 8 April 2021
Accepted: 18 May 2021
Published: 25 May 2021

Publisher's Note: MDPI stays neutral with regard to jurisdictional claims in published maps and institutional affiliations.

Copyright: © 2021 by the authors. Licensee MDPI, Basel, Switzerland. This article is an open access article distributed under the terms and conditions of the Creative Commons Attribution (CC BY) license (https://creativecommons.org/licenses/by/4.0/).

1. Introduction

Zirconium (Zr) and Zr-based alloys and oxides find widespread applications in many of the current emerging technological applications. Specifically, these materials are notable in structural engineering, electronics, optoelectronics, magneto-electronics, electrochemical, and energy-related technologies. Zr is affiliated with Group IV of the periodic table and demonstrates similar chemical and metallurgical properties to titanium. Zr can be identified as a refractory (melting point higher than transition metals), reactive, and corrosion-resistant metal. Zr and Ti are sister metals of unique contrast. Zr is ranked 19th in abundance among chemical elements existing in the Earth's top layer.

German chemist Martin H. Klaproth introduced zirconium in 1789. Its purification process using iodide crystals and producing a ductile metal was performed by Arkel and Boer from Eindhoven, Holland [1]. Several alloy production programs started in the early 1950. The development of zirconium arose from increasing demands for the nuclear industry. A combination of several appealing features, such as resistance to irradiation damage, good corrosion-oxidation resistance, transparency to thermal neutrons, and adequate mechanical properties, has given Zr and its alloys the most suitable structural materials in nuclear applications like cladding nuclear fuels. Moreover, its extensive applications in nuclear submarines and research on developing its applications in the chemical process industries have grown remarkably. Zirconium also has found promising medical applications, such as surgical tools and instruments and implants [2,3].

Zirconium alloys can be categorized based on the two primary applications: nuclear and nonnuclear. Alloys for nuclear applications are named Zr–1Sn–1Nb, Zr–2.5Nb, Zr-1Nb, Zircaloy-2, and Zircaloy-4. The nonnuclear grades are named Zr700, Zr702, Zr704, Zr705, Zr706, where their production proceeds with low alloying contents. Nuclear grades are virtually free of hafnium, making alloys stabilized against thermal neutrons [1]. Generally, alloying elements can be categorized into two groups of α-stabilizers and β-stabilizers, where tin, aluminum, and oxygen promote the α phase transformation. The Zr-rich end of the binary phase diagram of the α-stabilizer alloying elements demonstrates peritectoid reactions. Moreover, adding iron, nickel, chromium, niobium, molybdenum, and hydrogen stabilizes the β phase and represented binary diagrams exhibit eutectoid or monotectoid reactions in the Zr-rich end [3]. The α structure is a hexagonal close-packed lattice, and the β phase exhibits a body-centered cubic lattice. Figure 1 displays both α and β crystal structures along with their crystallographic information.

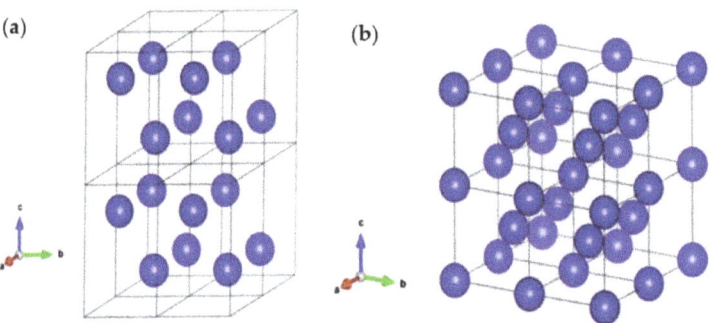

Figure 1. (a,b) crystal structures and crystallographic information of α and β phases in Zr [4].

Zircaloy-2 and Zircaloy-4 nominally contain 1.5% Sn, while the latter contains more iron but no nickel to hamper hydrogen evolution. Both alloys are used in water-controlled reactors, such as pressurized boiling water reactors. The Zr–2.5Nb alloy is implemented in heavy-water-controlled reactors. From the nonnuclear category, Zr702 is commercially pure Zr and the most well-known alloy for corrosion protection applications. Zr705 is the strongest alloy with enhanced formability. Zr700 and Zr706, with trace oxygen content, are good choices for several forming applications, such as explosive cladding and deep drawing [2].

Generally, zirconium needs surface modification to be suitable for its major applications. For biomedical applications, modifications need to provide appropriate functional characteristics on the surface that facilitate precipitation of hydroxyapatite compounds and integrate the implant with the surrounding environment. Several surface treatments, such as sol–gel deposition [5] and plasma spray [6], have been used for the purpose above. However, these techniques suffer from either non-crystallite oxide films or weak adherence films demanding post-treatment to attain crystalline coating with outstanding adherence [7]. Moreover, these treatments typically evoke cracking, spallation of the film, or alteration in the features of the Zr substrate.

Among the studied methods, PEO demonstrates superior features compared to other methods owing to its simplicity of processing and uncomplicated production of a porous oxide film with significant adherence to the surface of Zr [8]. Both chemical and morphological modifications contribute to the single-step processing of PEO coatings to boost the applicability of zirconium. Moreover, the PEO method offers various merits compared to other techniques, including a low-cost process, the feasibility to grow crystalline surface materials owing to the high-temperature process suitable for phase transformation, and easy control over morphology and thickness of the produced oxide layers by changing the electrolyte composition and PEO processing parameters [9,10].

The PEO of Zr and its alloys has drawn much attention for the last 15 years, indicating the success and importance of this method in a multipurpose surface modification. Compared to light metals (Ti, Al, and Mg), there is a remarkable deficiency in detailed information about the nucleation and growth mechanism, microstructure, morphology, various phases and their transformation, and effects of process parameters on coatings' properties that demand researchers to focus their studies in addressing them. From an application perspective, PEO studies can be categorized into two groups similar to alloy grades, namely nuclear power industry applications and biomedical applications.

This review aims to present the mechanism of oxide film formation and phase transformation in the PEO coating on Zr and its alloys. The microstructure and compositions of PEO coatings significantly influence the performance of a coating in different applications. We explain parameters' influence on the PEO coating and compare them for different applications. It should be noted that comprehensive identification of the growth mechanisms of the ceramic film using the PEO process on Zr-based substrates is currently not possible as the mechanisms are still obscure to some extent. Therefore, adopting models and theories already suggested for other valve metals like aluminum and magnesium requires us to fully discuss the discharge types and models after the breakdown voltage and finally to check the conformity of the reported surface morphology with the adopted mechanism.

2. The Growth Mechanism of PEO Coating: The Influence of Process Parameters

Generally, the PEO process on Zr is performed in either acidic or basic media. For the acidic media and galvanostatic conditions, it has been noted that the anodic films break down within the growing stage after the thickness of the oxide film has surpassed a critical value. This step is identified by the appearance of plasma discharges on the surface and a drop in the anodization rate (dE/dt). Eventually, the voltage fluctuates once the oxide film's destruction/rebuilding process occurs [11]. Figure 2a illustrates the anodization plots of Zr acquired galvanostatically at 20 mA/cm^2 in phosphate electrolyte at 5 °C. The anodization process can be divided into two recognizable steps: (I) the initial rise of the voltage (V), (II) the voltage fluctuations due to rapid formation and breakdown of the ZrO$_2$ film. The growth of the oxide film can be identified once the voltage rises linearly to reach the range of 300–350 V. At this step, ionic transport is the primary process controlling the film growth. A homogenous and compact film can be expected to grow on the Zr surface before the breakdown of the barrier layer [11]. While some slight deviation from linearity appears on the chronopotentiograph due to water oxidation on the oxide/electrolyte interface, voltage otherwise increases to the verge of a breakdown, where a distinct slope alteration occurs in the anodic graph. This event is identified by a series of simultaneous oscillations in anodization, crystallization of local oxide, and increasing internal lattice stress [12]. Once the voltage approaches the breakdown voltage (V_B), numerous micro sparks appear on the surface, indicating the beginning step of the PEO process. The chemical reactions occur between the participants of plasma and Zr electrodes in discharge channels [13]. The voltage fluctuations are related to the localized rupture/dissolving of the ZrO$_2$ layer followed by the quick reformation of the oxide in the same origin. Ikonopisov et al. also showed that increasing the current density makes the anodization rate faster in the acidic electrolyte. Monitoring current density, anodization time, and electrolyte composition confirm the importance of these factors on altering discharge regimes. The electrolyte composition and current density affect the discharge density. Moreover, the anodization time and electrolyte composition influence the average area and discharge lifetimes [14]. As shown in Figure 2b, Sandhyarani et al. categorized the PEO process after the voltage breakdown into three states, including the dynamic sparking, the near steady-state (arcing), and the steady-state PEO, which is above the critical voltage (V_C) [13]. During the dynamic sparking of the PEO process, the voltage rises at a lower rate compared to early anodic oxidation of the Zr substrate because small, dense microdischarges are created with the contribution of ionic species and electrons to the current [15]. The creation of abundant micro sparks circulating over the whole surface is the primary characteristic of this step. During the near steady-state of the PEO process, the rate of increasing voltage generally declines to a lower value than the previous step on the order of one-tenth. At the second step, the intensity of microdischarges circulating quickly throughout the surface rises remarkably. At the final state, the voltage inclines to a steady-state trend concerning time, and no further rise seems to occur. This steady-state behavior can be explained by considering that most of the total current derives from the electric current related to the dielectric breakdown and the constant resistance of the film. Furthermore, at the final state, strong concentric and individual discharges with amplified size and prolonged periods are created on the surface. Thus, the surface suffers from the nonuniform distribution of discharges, meaning some regions remain discharge-free.

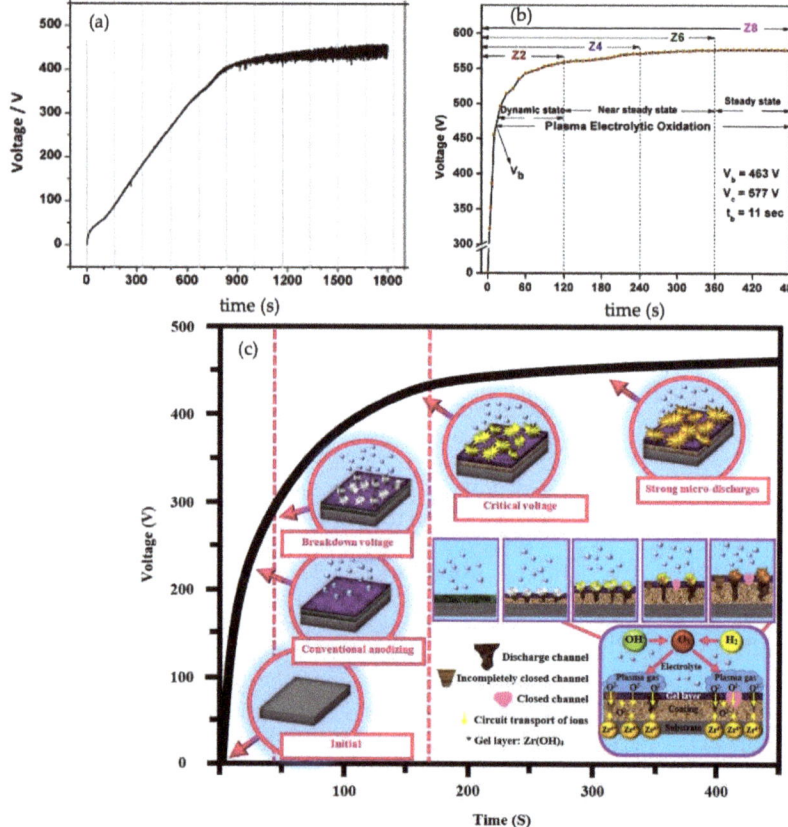

Figure 2. (a) Anodization curves of Zr at 5 °C obtained at 20 mA/cm^2 in 0.5 mol/L H$_3$PO$_4$ [11] (reproduced with permission number: 5034230474964, Elsevier); (b) chronopotentiograph for the PEO process at 0.150 A/cm^2 in 5 g/L Na$_3$PO$_4$·12H$_2$O electrolyte for 8 min [13] (reproduced with permission number: 5043471087024, Elsevier); (c) schematic demonstration of the PEO process as a function of time.

The whole PEO process displaying the development of microdischarges as a function of time is illustrated schematically in Figure 2c, which represents the mechanism of oxide film formation during the PEO process. As shown in Figure 2c, the PEO process on Zr can be performed at potentials above V_B. At these potentials, the remaining passive film cannot tolerate the strong electrical field and dielectric breakdown; therefore, it happens along with sparking over the metal surface [9,12,14]. Two phenomena are simultaneously involved in the PEO film growth on Zr: electrochemical reactions and plasma discharging processes over the metal/electrolyte interface [12]. In fact, plasma discharges are the main motive force for the growth of oxide layers, as they inject energy through numerous transient discharges that can melt the hitting surface, which is either metal or previously formed oxide film, and reconstruct the oxide film. Thus, fabrication of the PEO coating only demands a short treatment time to comprise a compact barrier layer and the components participating in the electrolyte. Unlike the direct current (DC) regime of the PEO process that the microdischarges are created only at the positive voltage side, the oxide film can form using both positive and negative waveform values under alternating current (AC) or pulsed bi-polar regime in excess of dielectric breakdown voltage. This condition promotes the formation of thick intermediate layers with a significant rise in hardness [16]. Arrabal et al. note that a thick intermediate layer can form during the PEO process using AC regime on Mg alloys after an abrupt decline in the voltage following the initial film growth at higher voltages [17]. Basically, the change in voltage after the initial step is associated with decreasing the intensity and appearance of the microdischarges, which are referred to as "soft sparking" [18]. Interestingly, the soft sparking phenomena were observed for the PEO process on Zr by Matykina et al. under AC regime and in the silicate electrolyte, where the voltage switched to reduced voltage after a treatment time of 18 min [19].

Chen et al. produced PEO coatings on Zircaloy-4 in silicate and pyrophosphate electrolyte separately and in their mixture for nuclear applications using an AC regime. They investigated the influence of the electrolyte composition on the coating features, including the microstructure and phase composition. It was noted that the steady-state voltage was the highest for the process in pyrophosphate electrolyte. In contrast, the process in silicate electrolyte showed the least [20]. Throughout the PEO process, microdischarges differ between aluminate, phosphate, and silicate electrolytes in terms of appearance and acoustic emission, and various coating compositions are stemmed from different microdischarge regimes and electrolyte composition.

The strong discharges initiating from the coating–substrate interface made the coating enriched in species derived from the substrate. In contrast, the near-surface discharges caused the participation of species from the electrolyte [20]. After the breakdown voltage, the initial microdischarges were short, plenty, and continuously disappearing and appearing at new locations for the process in silicate electrolyte. Therefore, the appearance of pancake-like features at the early stage of the PEO process could be attributed to forming of short-lived microdischarges and rapid solidification of molten materials in the discharge channels [20]. However, the solidification of molten materials to crystalline structure altered during the steady-state stage because fewer microdischarges were stronger and lasted longer at particular locations. Hussein et al. categorized discharge types originating from different regions during the PEO process primarily for the alloy of Al–Cu–Li [21–23]. We have suggested similar schematic representations for different types of discharges that might be created during the PEO process on Zr and its alloys. As shown in Figure 3a, discharges created during the PEO could be categorized into three types, including discharges initiating from the upper coating or gases and positioning over the coating surface (type A), discharges initiating from the substrate–coating interface (type B), and discharges originating inside pores and cracks in the coating (type C). Therefore, type A and C could explain the gaseous discharging conditions adjacent to the surface and inside the pores of the coating. In contrast, the strong discharge of type B originated from the metal/oxide film interface. Hussein et al. believed that the electrolyte species could incorporate into the coating layer through type A and C discharges. In contrast, type B mainly caused participation of species from the substrate [21]. Figure 3b illustrates the influence of type B discharges on the surface morphology of the PEO coating. Surface morphology and cross-section images are shown in Figure 3d,e to exemplify structural defects, pores, and cracks that may be created by type B discharges in the oxide film and adjacent to the coating/substrate interface. The appearance of type B discharges is associated with strong sparks. Moreover, this type of discharge promotes a pancake-like structure, which is common on the PEO coated surface of Zr alloys [20,24]. Later, a modified discharge model shown in Figure 3c introducing type D and E of discharges was suggested for Zr substrate based on the model for Al–Cu–Li alloys, suggested by Cheng et al. [24]. With a good conformability, the suggested model for the PEO coating on Zr in this review could explain the growth of the inner and outer layers. The internal pores adjacent to the inner layer/outer layer interface is the localized zone to receive the type D discharges. In contrast, the type E discharges strike the outer layer, creating large pores under pancake-like features, as shown in Figure 3c. Even though strong discharges of type D and E induced changes on the surface morphology, their impacts on the coating surface are less significant than that of the type B discharges. Therefore, this is the characteristic of microdischarges, either individually or collectively, that mainly dictates the phase formation, alteration of structures, and stress accumulation within oxide layers because of their main role in controlling the chemical and thermal conditions on the oxidizing surface. For this purpose, imaging techniques have been developed to study the microdischarge behavior during the PEO process. The spatial distribution, population density, size, and lifetime of the microdischarge could be recorded.

The application of optical emission spectroscopy (OES) to evaluate plasma discharges during the PEO process has recently drawn much attention among researchers. Cheng et al. captured the alteration of sparks' features during the PEO coating in the silicate electrolyte for 30 min [25]. They noted that the number of discharges reduced with treatment time, while their dimensions or intensities rose significantly. They attributed the reduction of the microdischarge number to the thickening effect of coatings, thus decreasing the number of weak sites. They finally categorized discharges during the PEO process based on their colors into three types. In the beginning, the discharges were white, small and numerous. In the second stage of the PEO process, between 3 and 10 min, larger sparks with diameters of ~0.28 mm appeared on the coating surface. In the last step of PEO (>10 min), the large and long-lasting sparks turned orange in hue on the surface [25]. Later, in a separate study, Cheng et al. investigated the role of silicate and aluminate electrolytes on the PEO behavior of Zircaloy-2 [26]. Figure 4a displays the relation between the PEO coatings' thickness and treatment time in aluminate, silicate, and phosphate electrolytes [26,27].

Zhang et al. investigated forming the PEO coating in aluminate electrolytes. They introduced paths for the progress of reaction during the formation of zirconia (ZrO_2) and alumina (Al_2O_3) phases [29]. In aluminate electrolyte, AlO_2^- ions could react with water and thus forming either $Al(OH)_4^-$ or $Al_n(OH)_{(4n+2)}^{(n+2)-}$ [30]. These negatively charged ions could react with Zr^{4+} and form ZrO_2-Al_2O_3 composite coatings. The order of reactions follows as [31,32]:

$$Zr \rightarrow Zr^{4+} + 4e^- \tag{1}$$

$$Zr^{4+} + 2OH^- + 2H_2O \rightarrow ZrO_2 + 2H_3O^+ \tag{2}$$

$$Zr^{4+} + AlO_2^- + 2H_2O \rightarrow ZrO_2 + 2Al_2O_3 \tag{3}$$

$$Zr^{4+} + Al(OH)_4^- \rightarrow ZrO_2 + Al_2O_3 + 2Al(OH)_3 + 5H_2O \tag{4}$$

$$4Al(OH)_4^- \rightarrow 2Al_2O_3 + 2OH^- + 3H_2O \tag{5}$$

After applying the voltage, a gas, mainly oxygen, evolved on the surface of the zirconium, thereby forming a thin layer of ZrO_2 on the substrate using reactions (1) and (2). After passing the breakdown voltage (i.e., 400 V), the oxide film broke, and many plasma discharges formed over the anode surface. Concurrently, Zr^{4+} metal cations diffused from the zirconium surface to the coating–electrolyte interface [31,32]. Zhang et al. deduced that the formation of Al_2O_3–ZrO_2 composite coatings was due to significant migration of AlO_2^- and $Al(OH)_4^-$ toward the anode in the prolonged PEO procedure under the electric field, and then deposition on the earlier formed layer [29].

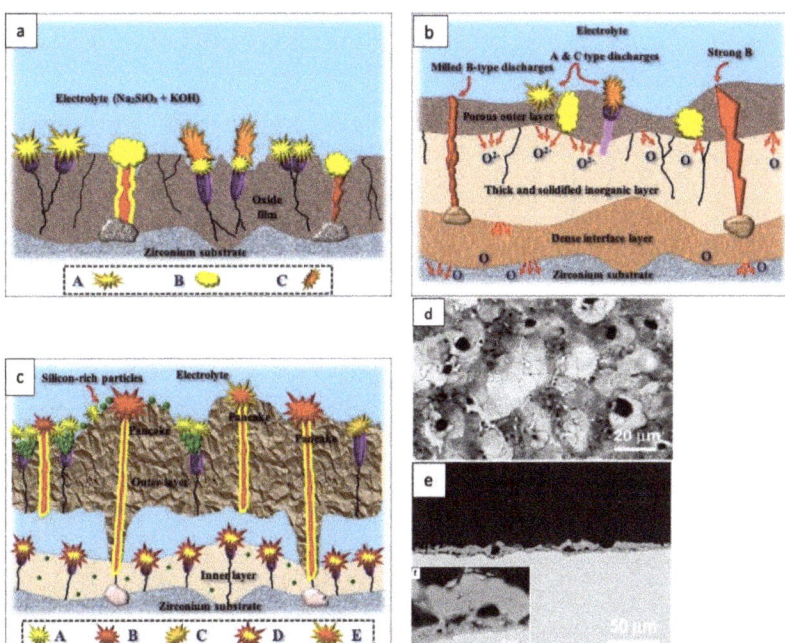

Figure 3. (a) Schematic illustration of the discharge types for the PEO coating on Zr in the silicate electrolyte based on the model suggested by Hussein et al. [21–23]; (b) Schematic representation of the discharge types for the PEO coating on Zr in the silicate electrolyte based on the model suggested by Cheng et al. [24]; (c) Effect of type B discharges on the surface morphology of the PEO coating on Zr substrate; (d,e) BS-SEM and cross-section image of the surface and the cross-section the PEO coating on Zircaloy-4 for 5 min at 100 Hz; (f) Magnified cross-section image from the coating shown in image (e) [28] (reproduced with permission number: 5043471384823, Elsevier).

It was noted that during the outward/inward growth mechanism of PEO coating, the inward growth had an important role in the entire growth of the oxide layer [34,35], indicating that the fresh layer close to the substrate was the result of inward growth. The outward coating was predominantly "annealed" in the electrolyte. This was why the aluminum in the depth of PEO coating was amorphous, but the region close to the surface contained the crystalized form of Al_2O_3. The "annealing" process induced abnormal growth of nanoplate-like α-Al_2O_3, while their deep growth was limited [29].

The appearance of discharges was recorded at various stages of the PEO treatment in both electrolytes. As shown in Figure 4d, the evolution of sparks' regime with time in the silicate electrolyte was similar to previous studies. However, this evolution did not behave similarly in the aluminate electrolyte after completing the early stage of the PEO process (600 s), shown in Figure 4e. After completing the PEO process, a striking contrast could be noticed between the surface morphologies of the two coatings from silicate and aluminate electrolytes. A uniform coating with a light appearance formed in silicate electrolyte throughout the whole treatment. However, the coating produced in aluminate electrolyte suffered blistering and spallation across the surface. The growth kinetics for these two coatings differed distinctively. A gradually accelerating rate of growth was recorded for coating formed in silicate electrolyte. In contrast, the growth rate in aluminate electrolytes was greater in the early coating stage (600 s). The rate then declined abruptly. The dissolution behavior of zirconium during the PEO in aluminate electrolyte could be divided into two regimes. First, during the pre-spallation stage of coating growth, the substrate initially underwent a relatively slow dissolution rate. After this, the dissolution rate switched to a much faster rate following the coating breakdown [26]. Sandhyarani et al. formed the PEO coating in electrolytes containing silicate, aluminate, and potassium hydroxide [36]. They reported that the electrolyte composition profoundly influenced the PEO voltage responses, such as the breakdown (V_B) and final voltage (V_f). Both V_B and V_f were higher in silicate electrolytes and decreased with KOH addition. Lu et al. also studied the effect of adding KOH and changing the duty cycle on the PEO process [37]. Adding KOH caused an abrupt drop in breakdown voltage from 500 to 264 V due to greater conductivity of electrolyte; however, they did not report the influence of duty cycle on alteration of the PEO process [37].

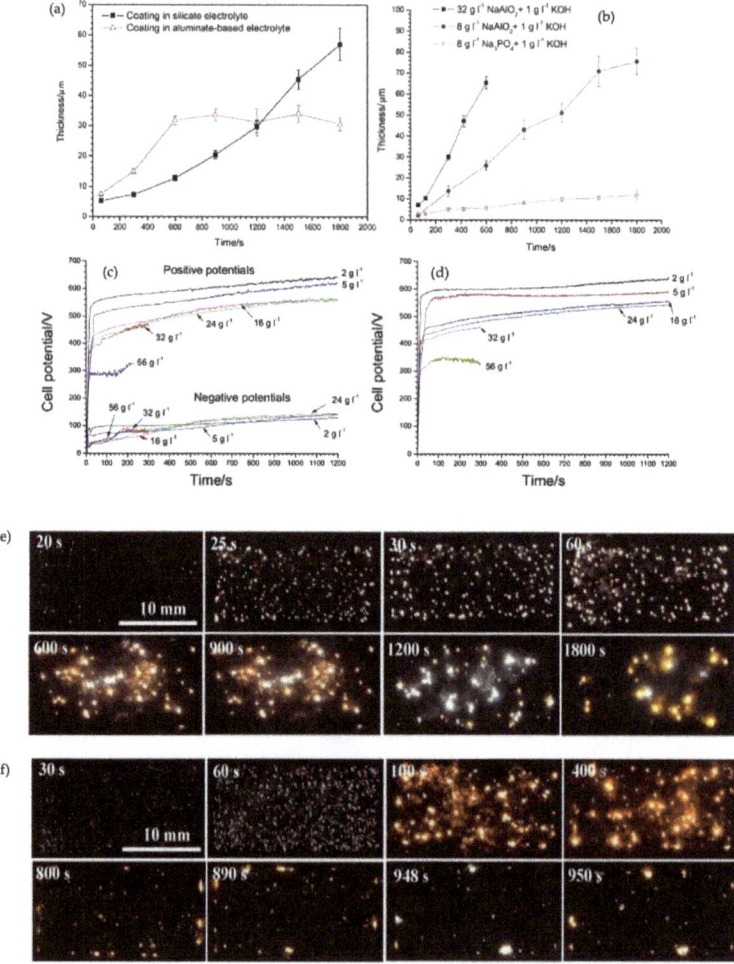

Figure 4. (a) Dependence of coating thickness on time of PEO on Zircaloy-2 in silicate and aluminate electrolytes and concentrations [26]; (b) Dependence of coating thickness on time of PEO on Zircaloy in different concentrations of aluminate electrolyte [27] (reproduced with permission number: 5043480179625 and 5043510569542, Elsevier); (c) and (d) Chronopotentiograph for the PEO process on Zirlo alloy in aluminate electrolyte with different concentration for pulse bipolar and pulse unipolar regimes, respectively [33] (reproduced with permission number: 5043511188587, Elsevier); (e) and (f) discharge appearances at different times of PEO in the silicate and aluminate electrolytes, respectively [26] (reproduced with permission number: 5043480179625, Elsevier).

Cheng et al. comprehensively studied the role of negative pulse and PEO coating formation mechanism on Zirlo alloy in aluminate electrolyte with different concentrations [33]. The plasma discharge regime influenced the PEO process extensively. As mentioned earlier, the discharge develops from fine sparks at the initial stage. It then shifts to stronger and more intensive sparks with less population during the steady-state step of oxide film thickening. Recently, a particular type of plasma discharges termed "soft sparking" have been identified in some PEO processes of valve metals [27,38–40]. Authors particularly examined the anion deposition process under different PEO conditions and associated alteration of the plasma discharge regime. Figure 4b,c demonstrated the chronopotentiograph under the bipolar and unipolar regimes, respectively [33]. It was evident that the breakdown voltage (V_B) dropped with aluminate concentration in electrolytes. This behavior could be attributed to the increased conductivity of the electrolyte, which could affect the overall cell potential [33,41].

3. Tailoring the Surface Morphology and Phase Composition of the PEO Coating

The morphology of the PEO coating is under the influence of electrolyte composition. Alteration of coatings' characteristics, such as compactness and the number of pores and cracks on the surface of coatings, can be justified in light of the electrical conductivity directly affected by the electrolyte composition.

The PEO coating by Yan and Han in 2007 was one of the earliest studies using a pure Zr disc and an aqueous solution containing calcium acetate monohydrate and β-glycerophosphate disodium salt pentahydrate (β-GPNa$_2$) in a dilute concentration [42]. Two years later, they used an alkaline electrolyte of α-GPNa$_2$ and glycerin to produce a PEO coating with greater biocompatibility. Simka and his coworkers investigated the influence of electrolyte concentration on the composition of the PEO coating. They reported a direct relationship between the concentration of the potassium silicate in the electrolyte and the silicon content of the PEO coating, where the Zr/Si atomic ratio decreased with increasing K$_2$SiO$_3$ concentration in the electrolyte [43]. Generally, treatment time directly influenced the PEO coatings' thickness. Increasing the thickness from 20 to 30 μm was reported for the PEO process on the pure Zr substrate once the oxidation time prolonged from 5 to 20 min [44].

The crystalline ZrO$_2$ contains three phases known as polymorphs under atmospheric pressure, as shown in Figure 5a–c. They are monoclinic (stable at temperatures below 1000 °C), tetragonal (stable at temperatures between 1000 and 1500 °C), and cubic (stable at temperatures above 1500 °C), where m-ZrO$_2$ is the most stable phase at the room temperature, and the other phases are stable at high-temperature [15]. Having said that, under equilibrium conditions, it has been seen that liquid ZrO$_2$ transforms to c-ZrO$_2$ at 2680 °C, then t-ZrO$_2$ appears at 2370 °C, and eventually, transformation ends up to m-ZrO$_2$ at 1240 °C [45]. For a short period of 10^{-3}–10^{-4} s, the local temperature rises to 10^4 K once the plasma forms on the surface [46]. Thus, the metallic Zr and its alloys transform to molten phase instantaneously under such high temperatures and solidify very quickly under non-equilibrium conditions. Therefore, this results in forming a mixture of monoclinic and tetragonal phases. The ratio of phases depends on how fast transformation occurs. The phase composition depends on several factors, including the electrolyte composition, the formation voltage, the surface pretreatment, the process treatment time, the alloying elements of the zirconium substrate, and additives participating in the PEO coating [20]. The coating properties are strongly influenced by the ratio of the zirconia phases present in the coating.

Similarly, a transition from tetragonal to monoclinic was reported for galvanostatic oxidation treatment. Accordingly, an impromptu phase transition occurs after the electrolyte breakdown process under the constant charge of 3.2 C/cm^2. At the initiation of the breakdown stage, both tetragonal and monoclinic phases coexist. The transition between the tetragonal and monoclinic phases happens because of the high strain energy related to the difference in the molar volume between the two phases [11]. The coexistence of both phases before an electrolytic breakdown can be explained in terms of the high field strength and the local energy created by the PEO process, confirmed by the intense spark generation. After the electrolytic breakdown in the galvanostatic regime of anodization, a continuous phase transformation proceeds from the semi-stable tetragonal phase to the stable monoclinic phase termed as a martensitic transformation.

Table 1 represents works performed by many researchers on the PEO coatings using various electrolytes and processing parameters.

Table 1. PEO process on Zr and Zr alloys representing different processing parameters.

Substrates	Electrolytes	PEO Processes Conditions				Coating Composition	Coating Features	Ref.
		Pulse Fre. (Hz)	Duty Cycle	Time (min)	App. Voltage or Current			
Pure Zr	0.2 M Ca(CH$_3$CO$_2$)$_2$·H$_2$O 0.02 M β-GPNa$_2$	100	30	5	350–500 V	Ca$_{0.15}$Zr$_{0.85}$O$_{1.85}$ (Ca-PSZ) m-ZrO$_2$	BS: 57.4 ± 2.1	[42]
Pure Zr	0.15 α-GPNa$_2$ 0.1 NaOH, 5 M Glycerin	100	26	5	400 V	m-ZrO$_2$ (major) t-ZrO$_2$ (minor)	Th.: 7	[17]
Pure Zr	K$_2$SiO$_3$, 5 g/dm^3 KOH	-	-	5	100, 200, 400 V	SiO$_2$, Zr$_2$SiO$_2$, ZrO$_2$	Th: 5 to 72	[14]
Pure Zr	0.05 M (H$_3$PO$_4$ or H$_2$C$_2$O$_4$)	Cons.	-	1 to 10	10 or 20 mA/cm^2	m-ZrO$_2$ and t-ZrO$_2$	-	-
Pure Zr	0.05 and 0.1 M (H$_3$PO$_4$ or H$_2$C$_2$O$_4$)	Cons.	-	Varies	3.2–43.2 C/cm^2	m-ZrO$_2$ and t-ZrO$_2$	-	[11]
Pure Zr	0.2–0.35 M Na$_2$AlO$_2$	100	26	30	400 V	α-Al$_2$O$_3$, t-ZrO$_2$, and m-ZrO$_2$	BS: 30–52	[18]
Pure Zr	5 g/L Na$_3$PO$_4$·12H$_2$O (TSOP)	50	95	2, 4, 6, 8	150 mA/cm^2	t-ZrO$_2$ (1–7 vol %) m-ZrO$_2$ (99–93 vol %)	Th: 3–17	[13]
Pure Zr	12 g/L Na$_2$SiO$_3$, 2 g/L KOH	50	-	5–120	480 (+), 120 (−) V 0.25 A/cm^2	m-ZrO$_2$ and t-ZrO$_2$	Th: 4.1–167	[45]
Zirlo^{TM}	30 g/L Na$_2$SiO$_3$·5H$_2$O and 2.8 g/L KOH	50	-	19 s, 30, 60	10 A/dm^2 150 V	c-ZrO$_2$, m-ZrO$_2$, t-ZrO$_2$	-	[19]
Zircaloy-4	30 g/L Na$_2$SiO$_3$·5H$_2$O + 4.88 g/L KOH 10 g/L Na$_4$P$_2$O$_7$·10H$_2$O 10 g/L (Na$_2$SiO$_3$·5H$_2$O + Na$_4$P$_2$O$_7$·10H$_2$O)	50	30	30	300 mA/cm^2	m-ZrO$_2$ and t-ZrO$_2$ or only m-ZrO$_2$	-	[20]
Zircaloy-4	30 g/L Na$_2$SiO$_3$·5H$_2$O + 4.88 g/L KOH	100	30	30	300 mA/cm^2	m-ZrO$_2$ and t-ZrO$_2$	Th: 15	[28]
Zircaloy-2	8 g/L Na$_2$SiO$_3$·9H$_2$O + 1 g/L KOH	1000	20	15 s to 30 min	400 mA/cm^2 (+), 300 mA/cm^2 (−)	m-ZrO$_2$ and t-ZrO$_2$	-	[25]
Zircaloy-2	8 g/L Na$_2$SiO$_3$·9H$_2$O + 1 g/L KOH 6 g/L NaAlO$_2$ + 8 g/L Na$_4$P$_2$O$_7$·10H$_2$O + 5 g/L KOH	1000	20	30	150 (+) mA/cm^2 100 (−) mA/cm^2	m-ZrO$_2$ and t-ZrO$_2$	-	[49]
Pure Zr	13 g/L (Na$_3$PO$_4$·12H$_2$O) 10 g/L (Na$_3$PO$_4$·12H$_2$O) + 3 g/L KOH 13 g/L (Na$_2$SiO$_3$·9H$_2$O) 10 g/L (Na$_2$SiO$_3$·9H$_2$O) + 3 g/L KOH 5 g/L (Na$_3$PO$_4$·12H$_2$O + Na$_2$SiO$_3$·9H$_2$O) + 3 g/L KOH	50	95	6	150 mA/cm^2	m-ZrO$_2$ (94), t-ZrO$_2$ (6) Vol % m-ZrO$_2$ (94), t-ZrO$_2$ (6) Vol % m-ZrO$_2$ (91), t-ZrO$_2$ (9) Vol % m-ZrO$_2$ (91), t-ZrO$_2$ (9) Vol % m-ZrO$_2$ (95), t-ZrO$_2$ (5) Vol %	Th: 95 ± 0.7 Th: 6.3 ± 0.4 Th: 7 ± 1.1 Th: 5.7 ± 0.9 Th: 6.9 ± 0.8	[46]
Zircaloy-2	8 g/L NaAlO$_2$ + 1 g/L KOH 32 g/L NaAlO$_2$ + 1 g/L KOH	1000	20	30 (dil.) 10 (con.)	150 mA/cm^2 (+), 100 mA/cm^2 (−)	γ-Al$_2$O$_3$, amorphous Al$_2$O$_3$ t-ZrO$_2$	Th: 78.5 Th: 65.4	[47]
Zircaloy-2	12 g/L Na$_2$SiO$_3$·9H$_2$O + 15 g/L (NaPO$_3$)$_6$	1000	8	20, 60	200 mA/cm^2 (+), 150 mA/cm^2 (−)	-	Th: 60, 122	[48]
Pure Zr	0.3 M NaAlO$_2$ and 0.03 M Na$_2$HPO$_4$	100	26	5, 15, 20, 30	400 V	Nanoplate-like α-Al$_2$O$_3$, m-ZrO$_2$, t-ZrO$_2$	-	[50]
Pure Zr	0.1 M C$_6$H$_8$O$_7$	-	-	1, 2, 3, 5, 10	100 mA/cm^2	m-ZrO$_2$	-	[51]
Pure Zr	30 g/L Ca(CH$_3$COO)$_2$·H$_2$O, 8 g/L C$_3$H$_7$Na$_2$O$_6$P·H$_2$O and 2 g/L KOH	400	-	5	4 A/cm^2	HA, Ca-PSZ, m-ZrO$_2$	-	[52]
Zirlo alloy	(2–56 g/L) NaAlO$_2$+1 g/L KOH	1000	20	5, 10, 20	0.14 (+) A/cm^2 0.05 (−) A/cm^2	(γ, α)-Al$_2$O$_3$, m-ZrO$_2$ and t-ZrO$_2$	-	[53]
Pure Zr	15 g/L NaSiO$_3$ + 2 g/L NaOH + 0.025 M AgC$_2$H$_3$O$_2$	-	-	3	400 (+) V 80 (−) V	ZrSiO$_4$, m-ZrO$_2$ and t-ZrO$_2$	-	[33]
Pure Zr	0.1 M K$_3$PO$_4$ + (0.01, 0.05, 0.1) M KOH	1000	25, 75, 100	-	200 mA/cm	(dominant) m-ZrO$_2$	-	[37]
Pure Zr	0.25 M Ca(CH$_3$CO$_2$)$_2$·H$_2$O + 0.06 M β-CaGP	50	-	5, 10, 15	0.292 A/cm^2	c-ZrO$_2$, CaZrO$_3$, HA	-	[54]
Pure Zr	TSOP: Na$_3$PO$_4$·12H$_2$O with (Al$_2$O$_3$, or CeO$_2$, or ZrO$_2$)	50	90	6	150 mA/cm^2	m-ZrO$_2$ and t-ZrO$_2$	5–7	[55]
Pure Zr	Ca(CH$_3$COO)$_2$·H$_2$O and C$_3$H$_7$Na$_2$O$_6$P·5H$_2$O	-	-	2.5–30	172 mA/cm^2	t-ZrO$_2$	-	[56]
Pure Zr	0.25 M Ca(CH$_3$COO)$_2$·H$_2$O + 0.06 M β-CaGP	-	-	15	260 mA/cm^2	c-ZrO$_2$, CaZrO$_3$, HA	-	[57]
Zircaloy-4	10 g/L Na$_2$SiO$_3$ + 3 g/L KOH 10 g/L Na$_3$PO$_4$ + 3 g/L KOH	600	8	20	550 V	m-ZrO$_2$ and t-ZrO$_2$	5 / 4	[58]
Pure Zr	0.25 M calcium acetate 0.06 M β-CaGP	-	-	5	0.370 A/cm^2	Ca$_{0.15}$Zr$_{0.85}$O$_{1.85}$ c-ZrO$_2$, CaZrO$_3$, Cu$_2$(P$_2$O$_7$)	21.8 ± 2.4	[59]
Zr alloy (Nb, Sn)	15 g/L Na$_2$SiO$_3$ + 3 g/L NaF + 15 g/L KOH with 0.1 g/L (Al$_2$O$_3$, MoS$_2$, CeO$_2$, and GO)	300	30	15	260 V	m-ZrO$_2$ and t-ZrO$_2$	-	[60]

(Th: Thickness vs. µm), (BS: Bond Strength vs. MPa).

The volume fraction of t-ZrO$_2$ and m-ZrO$_2$ can be estimated from the XRD pattern using the empirical equation suggested by Weimin et al. [61], as follows:

$$X_m = \frac{I_m(111) + I_m(\bar{1}11)}{I_m(111) + I_m(\bar{1}11) + I_t(111)} \times 100\% \qquad (6)$$

$$V_m = \frac{1.311 X_m}{1 + 0.311 X_m} \qquad (7)$$

$$V_t = 1 - V_m \qquad (8)$$

where X_m is the ratio of the total diffraction intensity of the major crystal planes in both tetragonal and monoclinic phases with volume fraction V_t and V_m, respectively, I_m is a preferential plane of the monoclinic phase, I_t is a preferential plane of the tetragonal phase. The transformation of tetragonal to monoclinic in zirconia is a reversible thermal martensitic transformation involving a significant temperature hysteresis (~200 °C), volume alteration (4–5%), and considerable shear strain (14–15%) [62].

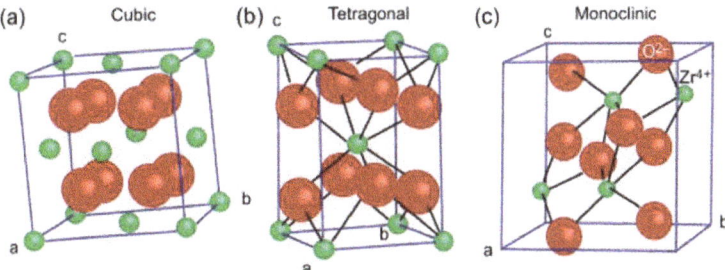

Figure 5. (a–c) Crystal structures and crystallographic information for monoclinic, tetragonal, and cubic phases in ZrO$_2$, respectively [63] (reproduced with permission number: 5055990208243, Elsevier).

Yan and Han reported a decreasing tendency to form m-ZrO$_2$ in the PEO coating with increasing the applied voltage [42]. The immediate temperature of the PEO discharge region varies between 10^3 and 10^6 °C during film formation. The transformation of m-ZrO$_2$ to c-ZrO$_2$ enhances once a high-temperature pulse is created with increasing the applied voltages [42]. Sandhyarani and his coworkers conducted the PEO coating in tri-sodium orthophosphate (TSOP) electrolyte for 2, 4, 6, and 8 min. They did not find separate phosphate phases in their characterizations, implying that the ZrO$_2$ film was doped with phosphorous (P). They also deduced that two factors influencing the phase transformation between monoclinic and tetragonal. The volume threshold of 7% was found for the tetragonal phase. Increasing the treatment time did not cause increased volume percentage of the tetragonal phase, and even the oxide film was produced at higher applied voltage. They inferred that the crystal size acted as an opposite factor for the higher localized temperature. Transformation above the threshold was restricted because crystals sintered to produce larger crystallites at higher temperatures induced a transformation from tetragonal to monoclinic with greater stability [13,64]. Authors reported that the PEO coatings produced in TSOP electrolyte with different treatment times demonstrated a smooth surface with very fine morphology, which was not similar to what was observed from other electrolytes [42]. They reported that discharge channels emerged on the whole surface in the form of circular spots spread out uniformly. Prolonging the PEO process caused a reduction of the population of discharge channels and enlargement of the discharge channel diameter, which could be ascribed to creating amplified sparks within the treatment time [13]. It is worth mentioning that prolonging the PEO treatment can result in stronger discharges amplified enough to sinter more elements grabbed from the electrolyte in the coating [7].

In another study, Yan et al. reported the formation of a porous surface with an average pore size of 3 µm, where the average pore size increased three times to offer greater feasibility for accumulation of hydroxyapatite compounds [47]. The PEO coating was composed of a major phase of m-ZrO_2 and a trace of t-ZrO_2, where a homogenous distribution of pores was found over the coating. The cross-section showed a continuous bilayer film, where the compact inner layer was 2 µm in thickness [47]. Simka et al. produced PEO coatings using different K_2SiO_3 concentrations of 0.1, 0.5, and 1 mol/dm^3 at a constant voltage of 400 V. They showed that the PEO coating produced at lower silicate concentration contained plenty of small pores with diameters less than 1 µm. The cross-section revealed that the coating was not uniform and that the outer layer of the coating separated in several regions [43]. Authors could show that the PEO coating was formed mainly at the metal–film interface owing to the inward migration of oxygen ions. Thus, the effect of the cation transport was not significant in the film growth. Moreover, silicon species could be traced at a relatively high concentration in the outer layer of the film with a thickness of approximately 20 nm. The formation of this exterior layer may be related to the outward transfer of Zr^{4+} ions, probably facilitated by the silica precipitation because of the reduction of the pH adjacent to the coating surface. The authors concluded that silicon could participate in the coating composition remarkably upon applying a potential above the breakdown voltage of the oxide layer. They also showed that silicon present in the coating in the form of silica and silicates.

Yan et al. produced Al_2O_3/ZrO_2 composite coatings using the PEO process in the electrolyte containing $NaAlO_2$ with different concentrations to systematically investigate the microstructure, bond strength, and microhardness of coatings [48]. The authors observed that increasing the concentration of $NaAlO_2$ changed the conductivity, which indirectly affected the anodization regime. At the fixed voltage, increasing conductivity was accompanied by the greater anodization current. Therefore, a more intensive plasma discharge would be created on the surface, where extremely intensive discharges were seen at 0.35 M $NaAlO_2$ inducing forming larger discharge pores. The morphology of the PEO coating formed at 0.2 M $NaAlO_2$ consisted of numerous nonuniformly distributed grains with different sizes and several cracks between accumulated grains. However, the needle-like crystals (α-Al_2O_3) emerged on the coating surface with increasing $NaAlO_2$ concentration. For this case, the coatings consisted of three layers: the outer layer, with the largest quantity of Al; the intermediate layer and the inner layer, where the gradian of Al content gradually shifted to zero at the coating/substrate interface [48]. The coating was dominantly composed of α-Al_2O_3, t-ZrO_2, and m-ZrO_2 phases. Finally, the authors understood that more α-Al_2O_3 and t-ZrO_2 phases developed in the coating composition with increasing $NaAlO_2$ concentration within a certain range.

Cengiz and Gencer reported the formation of PEO coatings on Zr samples in sodium silicate containing electrolyte and a coating process with different treatment times, as shown in Figure 6. First, the surface of the PEO coating produced in 5 min was evaluated. It was found that the surface was relatively smooth and contained the pancake-like features distributed irregularly with central micropores in the range of 1 µm. As displayed in Figure 6b–f, prolonging the treatment time decreased the number of micropores on the surface. At the same time, the pancake-like features rose significantly in their size to approximately 25 µm. As shown in Figure 6e,f, for the PEO coating produced above 20 min, the pancake-like features disappeared completely. A smooth surface appeared with a continuation of the coating process [45]. It was evident that, with the progression of the PEO process, a larger content of aggregated materials could accumulate adjacent to the opening of the pores and form irregular semispherical shapes with random distribution, as shown clearly in Figure 6d. Although most of the accumulated aggregates were dense, their structures were porous and distributed irregularly on the surface. It was found that increasing the number of microcracks and the appearance of flaking features on the surface resulted from the continuation of the PEO process for above 90 min [45]. It was also noticed that the surface roughness was impacted greatly by the treatment time, with roughness shifted from 0.63 to 8.3 μm as treatment time increased from 5 to 120 min. This tendency was also confirmed for the coating thickness, which rose linearly from 4.1 to 167 μm [45].

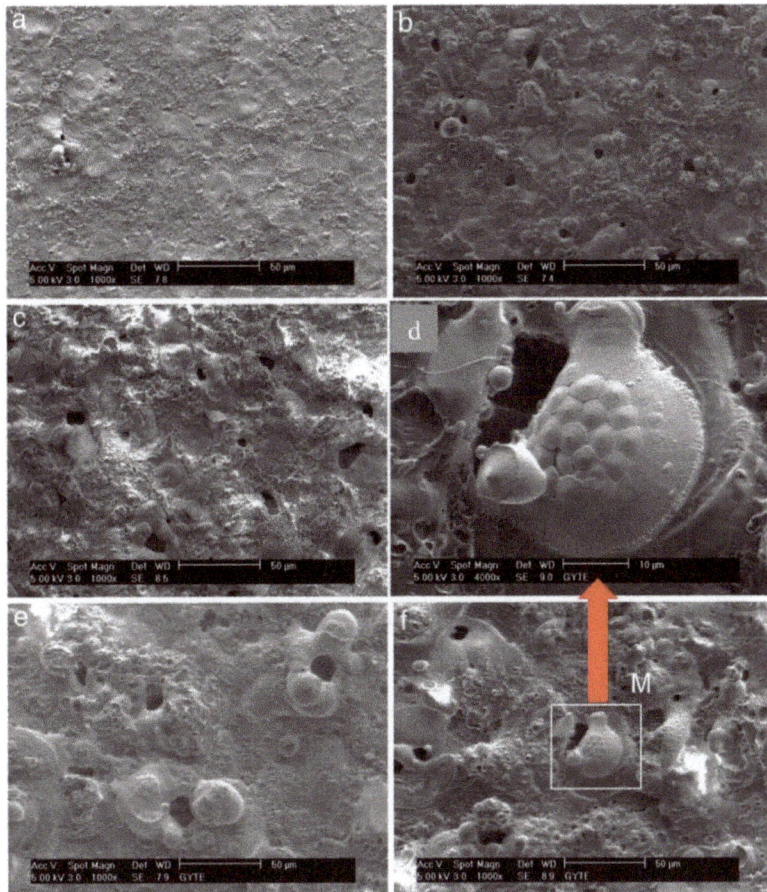

Figure 6. SEM images from the surface morphology of the PEO coating on pure Zr for the period of (**a**) 5, (**b**) 10, (**c**) 20, (**d**) 60, (**e**) 30, (**f**) 60 min. (**d**) High magnification of a typical SEM image of the cluster in equiaxed condition captured from the sub-figure (**f**) and marked as "M" for the processing time of 60 min [45] (reproduced with permission number: 5043520384143, Elsevier).

Another important remark was the composition of PEO coatings, where both m-ZrO_2 and t-ZrO_2 phases started to form from the beginning of the PEO coating process, and the dominant phase was monoclinic. In striking contrast, the authors found out that the phase content of the PEO coating formed in the silicate electrolyte was not significantly influenced by the coating thickness. Thus, they did not report any tendency in phase transformation resulting from the thickening of the coating, despite the results of other studies. In fact, the m-ZrO_2/t-ZrO_2 ratio remained constant during the PEO process. In contrast, due to a stabilizing effect of the silicate electrolyte, it was also reported that the t-ZrO_2/m-ZrO_2 ratio was higher in the outer layer compared to that in the inner layer of the coating [20]. This striking inconsistency can be ascribed to the presence of alloying elements.

Matykina et al. investigated the microstructure, morphology of the PEO coating on Zirlo[TM], a Zr alloy produced for nuclear applications [19]. They could show a three-layered coating produced for 30 min using the final voltage of 150 V. In fact, the PEO coating consisted of a thin, compact barrier layer formed anodically, an intermediate layer, and an outer layer rich in silicon. Authors found all three zirconia phases were present in the coating composition, where Si was also present in the top porous layer in the form of the amorphous compound. Moreover, they found the thickness of the outer layer did not change significantly with treatment time in the soft sparking region, where the elemental ratio of Si/Zr in the outer layer after the appearance of a soft sparking regime during the prolonged treatment was 10 times higher than that of the anodic film formed before the commencement of sparking [19].

In a comparative study on the role of electrolytes on the coating composition by Cheng et al., it was shown that the silicate electrolyte promoted the greater thickness of coating comprising outer layers enriched in silicon and inner layers consisting of m-ZrO_2 and t-ZrO_2 [20]. A remarkable feature on the coating surface produced in the silicate electrolyte at the early stages of striking microdischarges was forming the pancake-like features, implying extrusion of melted coating materials from discharge channels. However, the microstructure of the PEO coating formed during the steady-state process was an equiaxed dendritic center, in which the peripheral cellular orientation may be noticeable. In striking contrast, the PEO coating formed in the pyrophosphate electrolyte was much thinner and contained numerous cracks. The coating was only composed of m-ZrO_2, and the layered structure was not detected, unlike the coating formed in silicate electrolyte. The addition of silicate to the pyrophosphate electrolyte could improve the coating morphology by decreasing the number of cracks and boosting the proportionality of the tetragonal phase in the composition compared to the monoclinic phase. The flawed morphology observed for the coating formed in the pyrophosphate electrolyte could be attributed to the high stress created by the phase transformation from tetragonal to monoclinic. It was evident incorporating the silicon species from the electrolyte could partially stabilize the tetragonal phase. Cheng et al. studied the influence of different AC current frequencies on the morphology of the PEO coating in the silicate solution [28]. They reported the appearance of large pores spotted at the pancake-like features for the PEO coating produced at 100 Hz for 5 min on Zircaloy-4, indicating that the schematic model shown in Figure 3c was valid here and that sparks of type D had struck the surface. A fine dendritic structure was also observed in the outer region for the prolonged processing time.

After this, the authors pursued the PEO process by investigating the influence of sparking regimes on the phase composition and morphology of the coating produced on Zircaloy-2 [25]. They found the pancake-like feature prevailed the surface morphology of the PEO coating produced after 1 min. These features were formed because of type B discharges, illustrated schematically in Figure 3b. As mentioned, type B discharges were the strongest discharges during the PEO process originating from near the substrate/coating interface. Increasing treatment time caused enlarging dimensions of the pancake-like structures due to being struck by stronger discharges. Moreover, the detection of some pores on the dense outer layer surface and the porous inner layer could be ascribed to creating type D and type E discharges, as shown in the schematic model in Figure 3b. Moreover, the PEO coating composition altered with the silicate concentration in the electrolyte. The m-ZrO_2 dominated the coating in dilute silicate electrolyte, and t-ZrO_2 content increased at higher silicate concentration.

The microstructure of the PEO coating identifies with many factors, including the types and concentrations of electrolytes, the substrate composition and the electrical parameters. The surface morphology and cross-section images of the coating formed in silicate and aluminate electrolytes for 30 min are shown in Figure 7a–h, respectively. The surfaces in both coatings revealed pores, cracks, and pancake-like features, while the surface of the coating formed in aluminate was less rough than that formed in silicate. The typical features of solidification structures and spallation regions of the coatings were the two distinctive indicators of PEO coatings formed in silicate and aluminate electrolytes, as shown in Figure 7c,g, respectively [26]. Although the structure shown in Figure 7c was relatively rare, its creation required the presence of long-lasting molten materials due to prolonged sparks. The cross-sections illustrated in Figure 7d comprised a three-layer structure, including an inner barrier layer with a thickness of ~1 μm affixed on the alloy surface, an intermediate layer that was porous and cracked significantly, and an outer layer with extensive cracks. Figure 7f,g illustrated regions related to the coating's surface where part of the coating was separated due to spallation. The cross-section in Figure 7h displayed the morphology of the coating formed in aluminate electrolyte for 30 min and partially impacted by spallation. Later, Cheng et al. reported the formation of a bilayer PEO oxide in electrolytes containing $NaAlO_2$ and KOH for 30 min treatment [27]. It was also noted that coatings formed in aluminate electrolytes for shorter times resembled features similar to those shown in Figure 7e, such as pores, cracks, and pancake-like features. However, the spalled regions were absent. They found a noticeable connection between creating "soft sparking" and the inner layer growth. Dendritic growth of the tetragonal ZrO_2 phase throughout the outer layer was apparent, similar to other studies [20,25,26,28,65]. The thickness of the inner layer before and after the occurrence of "soft sparking" showed significant growth and thickening during this short period. The growth of the inner layer during the "soft sparking" period mainly filled the gap between the bilayer PEO coating [27]. Authors found that pancake-like features disappeared for the PEO coating formed in concentrated $NaAlO_2$ electrolyte, suggesting decreasing intensity and population density of sparks as the most effective factors [25,27]. Unlike the study by Yan et al. [48], the PEO coating obtained by Cheng et al. in a less concentrated electrolyte did not contain needle-like crystals of α-Al_2O_3. The coating was only composed of tetragonal and monoclinic zirconia phases, in which the tetragonal phase was found in greater quantity in the outer layer for the coating formed in aluminate electrolyte [26].

Sandhyarani et al. studied the role of electrolyte composition on the structure and morphology of the PEO coating. They used 5 different mixtures of Na_2SiO_3, Na_3PO_4, and KOH to find a suitable electrolyte system with optimum properties for biocompatibility applications [36]. It was evident that the relative proportion of tetragonal to monoclinic phases altered concerning electrolytes. Again, reducing the monoclinic phase in electrolytes containing silicate confirmed that Si stabilized the tetragonal phase at low temperatures. Sandhyarani and his coworker found that increasing silicate ion concentration caused alteration of preferential orientation from ($\bar{1}11$) to (200) in the monoclinic phase, in which the oxide film formed in a higher concentration of Na_2SiO_3 was grown fully in (200) orientation [36]. This was inconsistent with other studies, in which the growth of m-ZrO_2 film in (200) orientation was not reported for the PEO coating on Zr and its alloys in silicate electrolyte [19,25–27,43,45,49,65,66]. After considering studies on the PEO coating in silicate electrolyte, authors concluded that the alteration of orientation in oxide films was probably due to the duty cycle effect, in which switching from low duty cycles (26–30%) to high duty cycles (95%) significantly shifted the oxide film growth in (200) orientation. These phenomena could be explained by considering the role of high duty cycles on inducing a higher localized heat with the PEO treatment time and increasing the surface mobility of molten oxide that all eventually ended up changing to surface rearrangement of growing crystalline zirconia [36]. The addition of KOH to electrolyte modified the surface morphology significantly. The porosity and roughness of the oxide films decreased, and the coating became more uniform. Furthermore, PEO coating formed in silicate electrolyte demonstrated greater wettability and surface energy.

Zhang et al. produced Al_2O_3–ZrO_2 composite coatings at 400 V for treatment time from 5 to 30 min [29]. The PEO coating formed at prolonged treatment time contained the highest quantity of crystalline nanoplate-like alumina, distributed uniformly, interweaved with each other, and placed vertically over the surface [29]. It was remarkable that Zhang and his coworkers found the ratio of t-ZrO_2/m-ZrO_2 rose from depth to surface of the PEO coating formed in aluminate electrolyte.

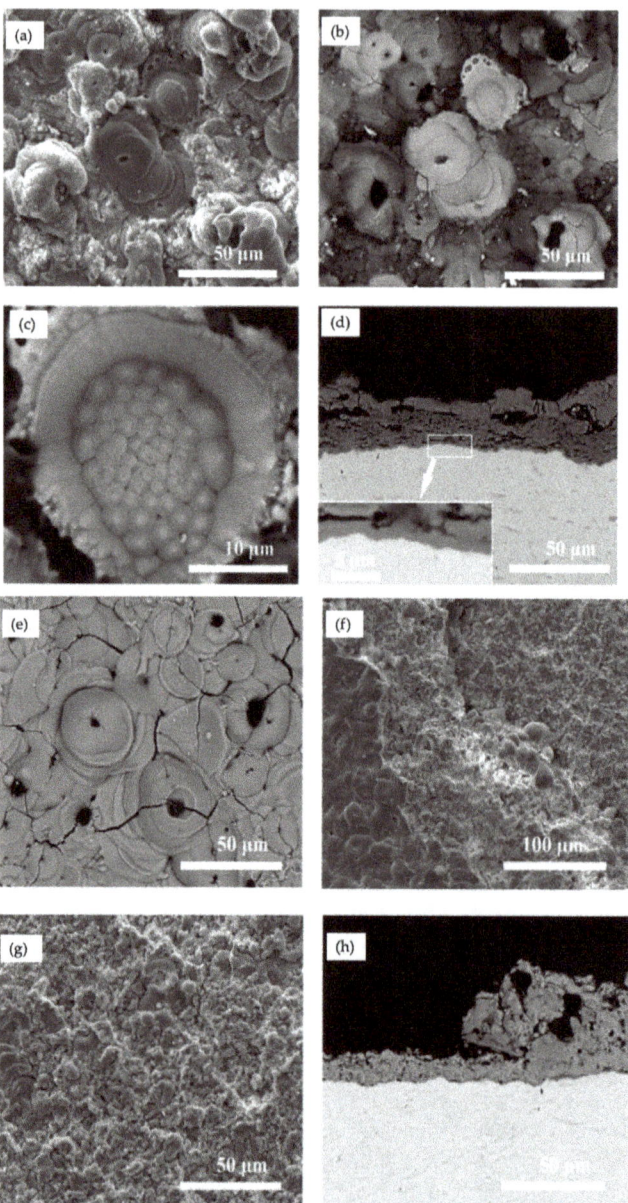

Figure 7. (**a–c**) Surface morphology, and (**d**) Cross-sectional image of the PEO coating formed for 1800 s in the silicate electrolyte; (**e–g**) Surface morphology, and (**h**) Cross-sectional image of a coating formed for 1800 s in the aluminate electrolyte [26] (reproduced with permission number: 5043480179625, Elsevier).

The effect of anion deposition in different concentrations of aluminate electrolyte was studied by Cheng et al. [41]. The surface morphology of coatings was examined for the entire various concentration of aluminates used in electrolytes. First, in dilute electrolyte with 2 g/L NaAlO$_2$, the pancake-like features were reported for both unipolar and bipolar conditions, in which each pancake contained a pinhole at the center due to the outlet of gases accumulated under the pancake-like features created by the strong discharges (type E) [10,38]. Figure 8 illustrates cross-sectional images of all PEO coatings formed at $NaAlO_2$ concentrations of 2, 16, 32, and 56 g/L. The coating thickness formed during the bipolar regime was very thin (~10 μm) and exhibited a single-layer nature in the cross-section shown in Figure 8a. However, the coating formed under a unipolar regime was a bi-layered structure with a thickness of ~30 μm. Remarkably, large lateral pores trans passing the compact outer layer and the thin inner layer were apparent for the coating formed in an electrolyte containing 2 g/L $NaAlO_2$, as shown in Figure 8b. The cross-sectional images for coating formed in 16 g/L $NaAlO_2$ under unipolar and bipolar regimes were different, as shown in Figure 8c,d. The unipolar structure resembled the structure formed in a dilute electrolyte, in which internal pores were detected. In contrast, the intervals between outer and inner layers in the bipolar coating were filled with dark coating materials instead of being void, as shown in the central part of Figure 8d. EDS analysis for areas shown in figures is shown in Figure 8. The ratio of Al/Zr highlighted the fact that the concentration of aluminate in electrolytes greatly influenced the proportion of anions incorporating into the coating. For coatings formed in an electrolyte containing 32 g/L $NaAlO_2$, the cross-section image of the coating formed under a unipolar regime illustrated the presence of big pores and generated interruption in the continuity of the coating, similar to coatings formed in dilute electrolytes. In striking contrast, the surface morphology of bipolar coating did not show pancake-like structures. At the same time, some white patches appeared on the surface of the bipolar coating. The coating growth behavior altered significantly for the PEO process is highly concentrated aluminate electrolyte (56 g/L $NaAlO_2$), in which the entire surface covered with patches of white materials deposited at a high rate. A prolonged process at highly concentrated electrolyte caused alteration of the morphology distinguishably. The pancake-like features disappeared from the coating surface. The surface of coatings contained noodle-like features. The surface morphology was extremely nonuniform for the unipolar regime, while mound-like features were created by a "sintering arc" being completely visible. The thickening of the coating under the unipolar regime, shown in cross-sectional images in Figure 8g, represented an extraordinary behavior due to protruding mound-like features. Moreover, the appearance of long cracks at the coating–substrate interface indicated the weak adherence of the coating. In contrast, the coating formed under the bipolar regime was more uniform. With no mound-like features appearing, the cross-section of coating indicated greater uniformity, and interfacial cracks did not progress [33].

Studying the mechanism of the PEO coating in different electrolyte concentrations taught us that pancake features were created on the surface during the process in low and moderate electrolyte concentrations under unipolar regimes. Pancake-like features were mainly created by discharge types B or E due to a dielectric breakdown process. Gas evolution accompanying the strong discharges generated large internal pores [38]. Figure 8j shows a suggested model by Cheng et al. for explaining the formation mechanism of pancake-like features and big pores [33]. Visually, each discharge spark was encapsulated by a non-luminous gas bubble [67]. It could be envisioned that negatively charged ions from electrolyte situated on the bubble surface and electron emissions from the electrolyte into the gas bubble stimulated discharges; however, electrons could be positioned at the oxide–electrolyte interface unrelated to the gas bubble [68]. To model the PEO process in highly concentrated electrolytes, the schematic presentation shown in Figure 8i could be suggested for the growth of the coating under a unipolar regime. The shown "soft sparking" was generated by heavy anion deposition and the accumulation of new compounds on the surface. The sintering arcs could generate the fast precipitations of alumina. Weak sparks were responsible for creating the conventional anodic film, making flower-like features on the surface.

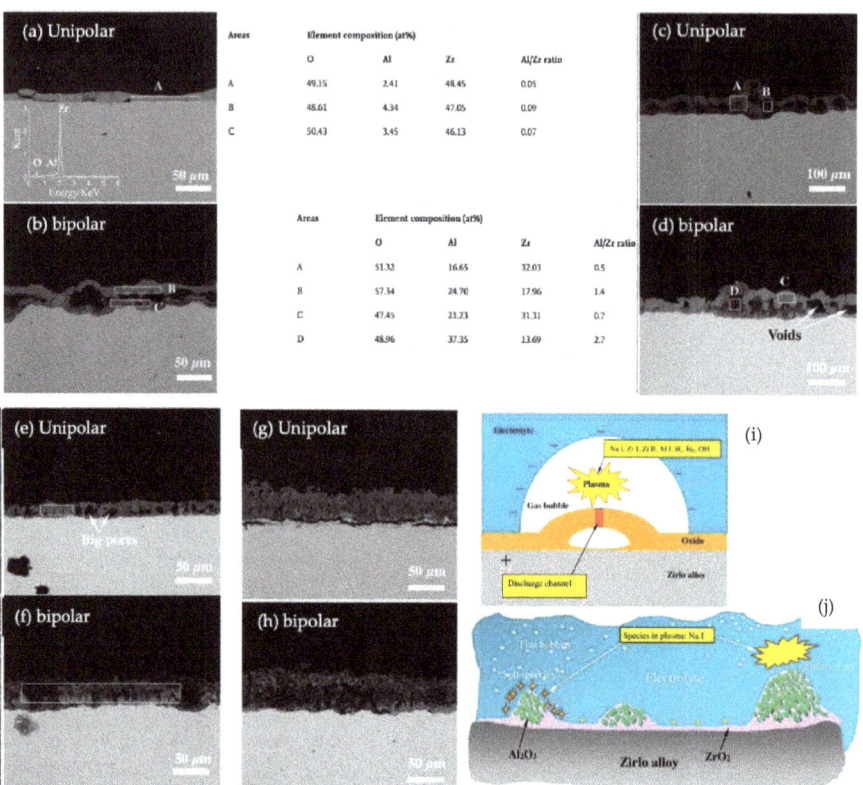

Figure 8. The cross-sectional images of the coatings under unipolar and bipolar regimes in electrolyte containing: (**a**,**b**) 2 g/L NaAlO$_2$ + 1 g/L KOH; (**c**,**d**) 16 g/L NaAlO$_2$ + 1 g/L KOH; (**e**,**f**) 32 g/L NaAlO$_2$ + 1 g/L KOH; (**g**,**h**) 56 g/L NaAlO$_2$ + 1 g/L KOH; (**i**) Schematic illustration of the growth steps through discharge penetration; (**j**) Schematic illustration of the coating formation process in highly concentrated electrolyte and under unipolar regime [33] (reproduced with permission number: 5043511188587, Elsevier).

Arun et al. used a single-step PEO procedure accompanied by electrophoretic deposition (EPD) to produce a composite coating in phosphate electrolyte by incorporating nanoparticles, including Al$_2$O$_3$, CeO$_2$, and ZrO$_2$, separately [55]. They studied microstructural alteration and found that the surface morphology was modified significantly after the involvement of nanoparticles. The participation of Al$_2$O$_3$ and CeO$_2$ nanoparticles promoted phase transformation from monoclinic to tetragonal in ZrO$_2$. At the same time, the surface nature turned to hydrophobic with the incorporation of particles due to less porous morphology. Another study also reported the incorporation of SiC into ZrO$_2$ composite coating using the EPD method in aluminate, phosphate, and silicate electrolytes. Monoclinic zirconia was the dominant phase in phosphate and silicate electrolytes, while tetragonal zirconia was dominant in aluminate electrolytes [69].

4. Concluding Remarks and Future Outlook

The ceramic oxide base coating can be synthesized on Zr and its alloys successfully using the PEO process in different alkaline and acidic electrolytes with outstanding coating adherence to the substrate. The wavy feature of the metal/coating interface could induce greater coating adherence to the substrate. The PEO coating depending on process parameters, could result in various surface morphologies, including various populations of pores and cracks, different surface roughness, different coating composition and features.

Generally, the thickness of the oxide film rose with the treatment time, consequently approaching the breakdown voltage. After this, there were three regions of discharge regimes, including dynamic discharges, near steady-state, and steady-state. Indeed, the number of discharges and discharge channels reduced with thickening the PEO coating. The discharge regimes could be categorized based on the electrolyte concentration, in which strong plasma discharges were dominant at dilute or moderately concentrated electrolyte, while the dominant discharges in concentrated electrolyte were sporadic patches of weak discharges or some localized "sintering arc".

Thus, discharge channels enlarged in size because stronger sparks struck the surface. This, in turn, led to increasing the content of debris materials thrown out from plasma channels and accumulated surrounding plasma channels. The ejaculation of materials through plasma channels was stimulated by high pressure that was created by evaporated materials. Hence, at stronger sparks' locations, more materials were ejaculated and splashed out of the plasma channels forming spots that look like geothermal boiling mud pools.

Using the silicate electrolyte for the PEO process promotes greater sintering of the outer layer through more significant participation of elements driven from the electrolyte and reducing the melting temperature of ZrO_2. The PEO coating formed on Zr alloys in the silicate electrolyte consisted of a three-layered coating, including a thin barrier layer, a porous inner layer, and a relatively compact outer layer. Both tetragonal and monoclinic phases were present in the inner and outer layers. An enriched silicon layer could be seen on the surface of the coating. However, plenty of cracks were found on the coating formed in the pyrophosphate, which can be attributed to phase transformation from tetragonal to monoclinic zirconia. The different morphologies of the coatings could be related to the differing natures of the microdischarges and participation of silicon species in the coatings. Different distribution of t-ZrO_2 and m-ZrO_2 between the outer layer and the inner layer of the PEO coating could be related to the different thermal conditions in the upper and lower coating regions.

The presence of silicon species in electrolytes has demonstrated its effectiveness in stabilizing the tetragonal phase, where the proportion of t-ZrO_2 in the coating composition intensified with increased concentration of the silicate in the electrolyte. For instance, in one case, t-ZrO_2 content rose from ~8 to 27 wt % by increasing the concentration of $Na_2SiO_3 \cdot 9H_2O$ from 8 to 30 g/L. However, the PEO coating in the phosphate-based electrolyte was dominantly composed of m-ZrO_2. PEO coatings grew at higher rates in concentrated aluminate electrolytes, while the intensity and population of discharges noticeably decreased. In concentrated electrolytes, pancake-like features disappeared, tetragonal zirconia became the dominant phase, and γ-Al_2O_3 started growing at the prolonged PEO procedures. Pancake-like features were created on the surface of PEO coating due to penetration of strong discharges generated either in dilute electrolytes for both bipolar and unipolar regimes or in moderately concentrated electrolytes for only unipolar regimes. In concentrated electrolytes, the dominant coating formation mechanism was the anion deposition, which changed the nature of PEO discharges.

The main intent of this review is to trigger inspiration for new research that will develop and expand the science and technique of the PEO coating fabrication with improved features and broader applications. To help provide such inspiration and provoke new research perspectives for future research in this exciting field, selective ideas for new research are presented in the following paragraphs. Despite the considerable research performed over the past three decades, there have been no definitive experiments that discernibly shed light on the mechanism of phase transformation from cubic to tetragonal and then to monoclinic. Many efforts have been devoted to identifying optimum process parameters to increase the proportionality of the tetragonal phase over the monoclinic phase. However, the role of each parameter, such as the composition of the electrolyte, electrical processing parameters, has not been comprehensively discovered. Clarification of the effect of quenching rate from tetragonal phase to monoclinic phase demands systematic experiments. Therefore, more work aimed at understanding the martensitic transition from tetragonal to monoclinic is warranted. We recommend experiments to be performed in the following areas:

Advanced surface characterization: The nature of oxide film in the form of layers stacking over the substrate and different ratios of tetragonal phase to monoclinic phase in different layers need to be addressed. Why does the tetragonal phase form in lower proportionality in the inner layer compared to the outer layer? Are strong discharges the only factor in forming a greater quantity of monoclinic phase in the inner layer?

In situ analysis: What is the temperature of the plasma region at the substrate–coating interface, where the inner layer forms? In situ analysis is complicated by the evolution of gases at the origination of discharges.

What is the optimum ratio of tetragonal to monoclinic phase to obtain the best performance in various applications? How can the aging effect be retarded to prevent the transition of tetragonal to monoclinic during prolonged use?

What is the role of post-treatments in modifying the surface features of PEO coatings? Heat treatment and laser treatment of surface and controlling the rate of quenching from higher temperature to room temperature are facile procedures to change the ratio of tetragonal to monoclinic phase. The ratio of tetragonal to monoclinic phases can be shifted to the same ratio for different stacking layers of the PEO coating. This modification can prevent the accumulation of mismatching stress between layers and prevent lateral detachment of layers.

Author Contributions: C.V.R. conceived and supervised this work. N.A. collected all the data, obtained permission to use the data existing in the literature, and compiled the scientific validations. Both authors involved in preparing the manuscript. Finally, All authors have read and agreed to the published version of the manuscript.

Funding: The authors also acknowledge, with pleasure, support from the National Science Foundation (NSF) with NSF-PREM grant #DMR-1827745.

Acknowledgments: N.A. acknowledges with pleasure the technical support and encouragement provided by the Center for Advanced Materials Research (CMR), UTEP. N.A. also acknowledges the Research Associate opportunity provided by CMR, UTEP.

Conflicts of Interest: The authors declare no conflict of interest.

References

1. Yau, T.-L.; Sutherlin, R.C.; Chang, A.W. Corrosion of Zirconium and Zirconium Alloy. In *Corrosion: Materials*; Stephen, D.C., Bernard, S.C., Eds.; ASM International: Ohio, OH, USA, 2018; Chapter 20; pp. 300–324.
2. Yau, T.-L.; Annamalai, V.E. *Corrosion of Zirconium and Its Alloys*; Elsevier: Amsterdam, The Netherlands, 2016; ISBN 978-0-12-803581-8.
3. Banerjee, S.; Banerjee, M. Nuclear applications: Zirconium alloys. *Ref. Modul. Mater. Sci. Mater. Eng.* **2016**. [CrossRef]
4. Xiao, B.; Sun, J.; Ruzsinszky, A.; Feng, J.; Haunschild, R.; Scuseria, G.E.; Perdew, J.P. Testing density functionals for structural phase transitions of solids under pressure: Si, SiO_2, and Zr. *Phys. Rev. B* **2013**, *88*, 184103. [CrossRef]
5. Uchida, M.; Kim, H.-M.; Kokubo, T.; Tanaka, K.; Nakamura, T. Structural dependence of apatite formation on zirconia gels in a simulated body fluid. *J. Ceram. Soc. Jpn.* **2002**, *110*, 710–715. [CrossRef]
6. Wang, G.; Liu, X.; Ding, C. Phase composition and in-vitro bioactivity of plasma sprayed calcia stabilized zirconia coatings. *Surf. Coat. Technol.* **2008**, *202*, 5824–5831. [CrossRef]
7. Sreekanth, D.; Rameshbabu, N.; Venkateswarlu, K. Effect of various additives on morphology and corrosion behavior of ceramic coatings developed on AZ31 magnesium alloy by plasma electrolytic oxidation. *Ceram. Int.* **2012**, *38*, 4607–4615. [CrossRef]
8. Pauporté, T.; Finne, J.; Kahn-Harari, A.; Lincot, D. Growth by plasma electrolysis of zirconium oxide films in the micrometer range. *Surf. Coat. Technol.* **2005**, *199*, 213–219. [CrossRef]
9. Mohedano, M.; Lu, X.; Matykina, E.; Blawert, C.; Arrabal, R.; Zheludkevich, M.L. *Plasma Electrolytic Oxidation (PEO) of Metals and Alloys*; Wandelt, K.B.T.-E., Ed.; Elsevier: Oxford, UK, 2018; pp. 423–438. ISBN 978-0-12-809894-3.
10. Kaseem, M.; Fatimah, S.; Nashrah, N.; Ko, Y.G. Recent progress in surface modification of metals coated by plasma electrolytic oxidation: Principle, structure, and performance. *Prog. Mater. Sci.* **2021**, *117*, 100735. [CrossRef]
11. Ikonopisov, S.; Girginov, A.; Machkova, M. Electrical breaking down of barrier anodic films during their formation. *Electrochim. Acta* **1979**, *24*, 451–456. [CrossRef]
12. Santos, J.S.; Lemos, S.G.; Goncalves, W.N.; Bruno, O.M.; Pereira, E. Characterization of electrical discharges during spark anodization of zirconium in different electrolytes. *Electrochim. Acta* **2014**, *130*, 477–487. [CrossRef]

13. Sandhyarani, M.; Rameshbabu, N.; Venkateswarlu, K.; Sreekanth, D.; Subrahmanyam, C. Surface morphology, corrosion resistance and in vitro bioactivity of P containing ZrO_2 films formed on Zr by plasma electrolytic oxidation. *J. Alloys Compd.* **2013**, *553*, 324–332. [CrossRef]
14. Ikonopisov, S. Theory of electrical breakdown during formation of barrier anodic films. *Electrochim. Acta* **1977**, *22*, 1077–1082. [CrossRef]
15. Venkateswarlu, K.; Rameshbabu, N.; Sreekanth, D.; Bose, A.C.; Muthupandi, V.; Babu, N.K.; Subramanian, S. Role of electrolyte additives on in-vitro electrochemical behavior of micro arc oxidized titania films on Cp Ti. *Appl. Surf. Sci.* **2012**, *258*, 6853–6863. [CrossRef]
16. Yerokhin, A.; Snizhko, L.; Gurevina, N.; Leyland, A.; Pilkington, A.; Matthews, A. Spatial characteristics of discharge phenomena in plasma electrolytic oxidation of aluminium alloy. *Surf. Coat. Technol.* **2004**, *177–178*, 779–783. [CrossRef]
17. Arrabal, R.; Matykina, E.; Hashimoto, T.; Skeldon, P.; Thompson, G. Characterization of AC PEO coatings on magnesium alloys. *Surf. Coat. Technol.* **2009**, *203*, 2207–2220. [CrossRef]
18. Jaspard-Mécuson, F.; Czerwiec, T.; Henrion, G.; Belmonte, T.; Dujardin, L.; Viola, A.; Beauvir, J. Tailored aluminium oxide layers by bipolar current adjustment in the Plasma Electrolytic Oxidation (PEO) process. *Surf. Coat. Technol.* **2007**, *201*, 8677–8682. [CrossRef]
19. Matykina, E.; Arrabal, R.; Skeldon, P.; Thompson, G.; Wang, P.; Wood, P. Plasma electrolytic oxidation of a zirconium alloy under AC conditions. *Surf. Coat. Technol.* **2010**, *204*, 2142–2151. [CrossRef]
20. Cheng, Y.; Matykina, E.; Skeldon, P.; Thompson, G. Characterization of plasma electrolytic oxidation coatings on Zircaloy-4 formed in different electrolytes with AC current regime. *Electrochim. Acta* **2011**, *56*, 8467–8476. [CrossRef]
21. Hussein, R.O.; Nie, X.; Northwood, D.O.; Yerokhin, A.; Matthews, A. Spectroscopic study of electrolytic plasma and discharging behaviour during the plasma electrolytic oxidation (PEO) process. *J. Phys. D Appl. Phys.* **2010**, *43*, 43. [CrossRef]
22. Hussein, R.; Nie, X.; Northwood, D. An investigation of ceramic coating growth mechanisms in plasma electrolytic oxidation (PEO) processing. *Electrochim. Acta* **2013**, *112*, 111–119. [CrossRef]
23. Hussein, R.O.; Northwood, D.O.; Nie, X. Processing-microstructure relationships in the plasma electrolytic oxidation (PEO) coating of a magnesium alloy. *Mater. Sci. Appl.* **2014**, *5*, 124–139. [CrossRef]
24. Cheng, Y.-L.; Xue, Z.-G.; Wang, Q.; Wu, X.-Q.; Matykina, E.; Skeldon, P.; Thompson, G. New findings on properties of plasma electrolytic oxidation coatings from study of an Al–Cu–Li alloy. *Electrochim. Acta* **2013**, *107*, 358–378. [CrossRef]
25. Cheng, Y.; Wu, F.; Matykina, E.; Skeldon, P.; Thompson, G. The influences of microdischarge types and silicate on the morphologies and phase compositions of plasma electrolytic oxidation coatings on Zircaloy-2. *Corros. Sci.* **2012**, *59*, 307–315. [CrossRef]
26. Cheng, Y.; Wu, F.; Dong, J.; Wu, X.; Xue, Z.; Matykina, E.; Skeldon, P.; Thompson, G. Comparison of plasma electrolytic oxidation of zirconium alloy in silicate- and aluminate-based electrolytes and wear properties of the resulting coatings. *Electrochim. Acta* **2012**, *85*, 25–32. [CrossRef]
27. Cheng, Y.; Cao, J.; Peng, Z.; Wang, Q.; Matykina, E.; Skeldon, P.; Thompson, G. Wear-resistant coatings formed on Zircaloy-2 by plasma electrolytic oxidation in sodium aluminate electrolytes. *Electrochim. Acta* **2014**, *116*, 453–466. [CrossRef]
28. Cheng, Y.; Matykina, E.; Arrabal, R.; Skeldon, P.; Thompson, G. Plasma electrolytic oxidation and corrosion protection of Zircaloy-4. *Surf. Coat. Technol.* **2012**, *206*, 3230–3239. [CrossRef]
29. Zhang, L.; Zhang, W.; Han, Y.; Tang, W. A nanoplate-like α-Al_2O_3 out-layered Al_2O_3-ZrO_2 coating fabricated by micro-arc oxidation for hip joint prosthesis. *Appl. Surf. Sci.* **2016**, *361*, 141–149. [CrossRef]
30. Martini, C.; Ceschini, L.; Tarterini, F.; Paillard, J.; Curran, J. PEO layers obtained from mixed aluminate–phosphate baths on Ti-6Al-4V: Dry sliding behaviour and influence of a PTFE topcoat. *Wear* **2010**, *269*, 747–756. [CrossRef]
31. Yan, Y.; Han, Y.; Huang, J. Formation of Al_2O_3–ZrO_2 composite coating on zirconium by micro-arc oxidation. *Scr. Mater.* **2008**, *59*, 203–206. [CrossRef]
32. Yerokhin, A.; Leyland, A.; Matthews, A. Kinetic aspects of aluminium titanate layer formation on titanium alloys by plasma electrolytic oxidation. *Appl. Surf. Sci.* **2002**, *200*, 172–184. [CrossRef]
33. Cheng, Y.; Wang, T.; Li, S.; Cheng, Y.; Cao, J.; Xie, H. The effects of anion deposition and negative pulse on the behaviours of plasma electrolytic oxidation (PEO)—A systematic study of the PEO of a Zirlo alloy in aluminate electrolytes. *Electrochim. Acta* **2017**, *225*, 47–68. [CrossRef]
34. Wang, C.; Wang, F.; Han, Y. Structural characteristics and outward–inward growth behavior of tantalum oxide coatings on tantalum by micro-arc oxidation. *Surf. Coat. Technol.* **2013**, *214*, 110–116. [CrossRef]
35. Li, J.; Cai, H.; Xue, X.; Jiang, B. The outward–inward growth behavior of microarc oxidation coatings in phosphate and silicate solution. *Mater. Lett.* **2010**, *64*, 2102–2104. [CrossRef]
36. Sandhyarani, M.; Prasradrao, T.; Rameshbabu, N. Role of electrolyte composition on structural, morphological and in-vitro biological properties of plasma electrolytic oxidation films formed on zirconium. *Appl. Surf. Sci.* **2014**, *317*, 198–209. [CrossRef]
37. Lu, S.-F.; Lou, B.-S.; Yang, Y.-C.; Wu, P.-S.; Chung, R.-J.; Lee, J.-W. Effects of duty cycle and electrolyte concentration on the microstructure and biocompatibility of plasma electrolytic oxidation treatment on zirconium metal. *Thin Solid Films* **2015**, *596*, 87–93. [CrossRef]
38. Cheng, Y.; Cao, J.; Mao, M.; Xie, H.; Skeldon, P. Key factors determining the development of two morphologies of plasma electrolytic coatings on an Al–Cu–Li alloy in aluminate electrolytes. *Surf. Coat. Technol.* **2016**, *291*, 239–249. [CrossRef]

39. Mécuson, F.; Czerwiec, T.; Belmonte, T.; Dujardin, L.; Viola, A.; Henrion, G. Diagnostics of an electrolytic microarc process for aluminium alloy oxidation. *Surf. Coat. Technol.* **2005**, *200*, 804–808. [CrossRef]
40. Matykina, E.; Arrabal, R.; Scurr, D.; Baron, A.; Skeldon, P.; Thompson, G. Investigation of the mechanism of plasma electrolytic oxidation of aluminium using 18O tracer. *Corros. Sci.* **2010**, *52*, 1070–1076. [CrossRef]
41. Cheng, Y.-L.; Cao, J.-H.; Mao, M.-K.; Peng, Z.-M.; Skeldon, M.-K.M.P.; Thompson, G. High growth rate, wear resistant coatings on an Al–Cu–Li alloy by plasma electrolytic oxidation in concentrated aluminate electrolytes. *Surf. Coat. Technol.* **2015**, *269*, 74–82. [CrossRef]
42. Yan, Y.; Han, Y. Structure and bioactivity of micro-arc oxidized zirconia films. *Surf. Coat. Technol.* **2007**, *201*, 5692–5695. [CrossRef]
43. Simka, W.; Sowa, M.; Socha, R.P.; Maciej, A.; Michalska, J. Anodic oxidation of zirconium in silicate solutions. *Electrochim. Acta* **2013**, *104*, 518–525. [CrossRef]
44. Lan, R.; Dong, L.; Wang, C.; Liang, T.; Tian, J. Influence of oxidation time on microstructure and composition of micro-arc oxidation coatings formed on zirconium. *Mater. Res. Innov.* **2014**, *18*, S2-123–S2-127. [CrossRef]
45. Cengiz, S.; Gencer, Y. The characterization of the oxide based coating synthesized on pure zirconium by plasma electrolytic oxidation. *Surf. Coat. Technol.* **2014**, *242*, 132–140. [CrossRef]
46. Dunleavy, C.; Golosnoy, I.; Curran, J.; Clyne, T. Characterisation of discharge events during plasma electrolytic oxidation. *Surf. Coat. Technol.* **2009**, *203*, 3410–3419. [CrossRef]
47. Han, Y.; Yan, Y.; Lu, C. Ultraviolet-enhanced bioactivity of ZrO_2 films prepared by micro-arc oxidation. *Thin Solid Films* **2009**, *517*, 1577–1581. [CrossRef]
48. Yan, Y.; Han, Y.; Li, D.; Huang, J.; Lian, Q. Effect of $NaAlO_2$ concentrations on microstructure and corrosion resistance of Al_2O_3/ZrO_2 coatings formed on zirconium by micro-arc oxidation. *Appl. Surf. Sci.* **2010**, *256*, 6359–6366. [CrossRef]
49. Zou, Z.; Xue, W.; Jia, X.; Du, J.; Wang, R.; Weng, L. Effect of voltage on properties of microarc oxidation films prepared in phosphate electrolyte on Zr–1Nb alloy. *Surf. Coat. Technol.* **2013**, *222*, 62–67. [CrossRef]
50. Cheng, Y.; Peng, Z.; Wu, X.; Cao, J.; Skeldon, P.; Thompson, G. A comparison of plasma electrolytic oxidation of Ti-6Al-4V and Zircaloy-2 alloys in a silicate-hexametaphosphate electrolyte. *Electrochim. Acta* **2015**, *165*, 301–313. [CrossRef]
51. Stojadinović, S.; Vasilić, R.; Radić, N.; Grbić, B. Zirconia films formed by plasma electrolytic oxidation: Photoluminescent and photocatalytic properties. *Opt. Mater.* **2015**, *40*, 20–25. [CrossRef]
52. Cengiz, S.; Uzunoglu, A.; Stanciu, L.; Tarakci, M.; Gencer, Y. Direct fabrication of crystalline hydroxyapatite coating on zirconium by single-step plasma electrolytic oxidation process. *Surf. Coat. Technol.* **2016**, *301*, 74–79. [CrossRef]
53. Fidan, S.; Muhaffel, F.; Riool, M.; Cempura, G.; De Boer, L.; Zaat, S.; Czyrska-Filemonowicz, C.; Cimenoglu, H. Fabrication of oxide layer on zirconium by micro-arc oxidation: Structural and antimicrobial characteristics. *Mater. Sci. Eng. C* **2017**, *71*, 565–569. [CrossRef]
54. Aktuğ, S.L.; Durdu, S.; Yalçın, E.; Çavuşoğlu, K.; Usta, M. In vitro properties of bioceramic coatings produced on zirconium by plasma electrolytic oxidation. *Surf. Coat. Technol.* **2017**, *324*, 129–139. [CrossRef]
55. Arun, S.; Arunnellaiappan, T.; Rameshbabu, N. Fabrication of the nanoparticle incorporated PEO coating on commercially pure zirconium and its corrosion resistance. *Surf. Coat. Technol.* **2016**, *305*, 264–273. [CrossRef]
56. Cengiz, S.; Azakli, Y.; Tarakci, M.; Stanciu, L.; Gencer, Y. Microarc oxidation discharge types and bio properties of the coating synthesized on zirconium. *Mater. Sci. Eng. C* **2017**, *77*, 374–383. [CrossRef] [PubMed]
57. Sandhyarani, M.; Rameshbabu, N.; Venkateswarlu, K.; Rama Krishna, L. Fabrication, characterization and in-vitro evaluation of nanostructured zirconia/hydroxyapatite composite film on zirconium. *Surf. Coat. Technol.* **2014**, *238*, 58–67. [CrossRef]
58. Wang, Y.; Feng, W.; Xing, Y.; Ge, Y.; Guo, L.; Ouyang, J.; Jia, D.; Zhou, Y. Degradation and structure evolution in corrosive LiOH solution of microarc oxidation coated Zircaloy-4 alloy in silicate and phosphate electrolytes. *Appl. Surf. Sci.* **2018**, *431*, 2–12. [CrossRef]
59. Aktug, S.L.; Durdu, S.; Aktas, S.; Yalcin, E.; Usta, M. Characterization and investigation of in vitro properties of antibacterial copper deposited on bioactive ZrO_2 coatings on zirconium. *Thin Solid Films* **2019**, *681*, 69–77. [CrossRef]
60. Li, Z.-Y.; Cai, Z.-B.; Cui, X.-J.; Liu, R.-R.; Yang, Z.-B.; Zhu, M.-H. Influence of nanoparticle additions on structure and fretting corrosion behavior of micro-arc oxidation coatings on zirconium alloy. *Surf. Coat. Technol.* **2021**, *410*, 126949. [CrossRef]
61. Ma, W.; Wen, L.; Guan, R.; Sun, X.; Li, X. Sintering densification, microstructure and transformation behavior of $Al_2O_3/ZrO_2(Y_2O_3)$ composites. *Mater. Sci. Eng. A* **2008**, *477*, 100–106. [CrossRef]
62. Basu, B.; Balani, K. Toughness optimization in zirconia-based ceramics. In *Advanced Structural Ceramics*; John Wiley & Sons: Hoboken, NJ, USA, 2011; pp. 173–214.
63. Farid, S.B. Structure, microstructure, and properties of bioceramics. In *Bioceramics: For Materials Science and Engineering*; Elsevier BV: Amsterdam, The Netherlands, 2019; pp. 39–76.
64. Mercera, P.D.L.; Van Ommen, J.G.; Doesburg, E.B.M.; Burggraaf, A.J.; Ross, J.R.H. Zirconia as a support for catalysts: Evolution of the texture and structure on calcination in air. *Appl. Catal.* **1990**, *57*, 127–148. [CrossRef]
65. Cheng, Y.-L.; Wu, F. Plasma electrolytic oxidation of zircaloy-4 alloy with DC regime and properties of coatings. *Trans. Nonferrous Met. Soc. China* **2012**, *22*, 1638–1646. [CrossRef]
66. Chen, Y.; Nie, X.; Northwood, D. Investigation of Plasma Electrolytic Oxidation (PEO) coatings on a Zr–2.5Nb alloy using high temperature/pressure autoclave and tribological tests. *Surf. Coat. Technol.* **2010**, *205*, 1774–1782. [CrossRef]

67. Troughton, S.C.; Nomine, A.; Henrion, G.; Clyne, T.W. Synchronised electrical monitoring and high speed video of bubble growth associated with individual discharges during plasma electrolytic oxidation. *Appl. Surf. Sci.* **2015**, *359*, 405–411. [CrossRef]
68. Wang, L.; Chen, L.; Yan, Z.; Fu, W. Optical emission spectroscopy studies of discharge mechanism and plasma characteristics during plasma electrolytic oxidation of magnesium in different electrolytes. *Surf. Coat. Technol.* **2010**, *205*, 1651–1658. [CrossRef]
69. Sukumaran, A.; Sampatirao, H.; Balasubramanian, R.; Parfenov, E.; Mukaeva, V.; Nagumothu, R. Formation of ZrO_2–SiC composite coating on zirconium by plasma electrolytic oxidation in different electrolyte systems comprising of SiC nanoparticles. *Trans. Indian Inst. Met.* **2018**, *71*, 1699–1713. [CrossRef]

Review

Plasma Electrolytic Oxidation Ceramic Coatings on Zirconium (Zr) and Zr-Alloys: Part-II: Properties and Applications

Navid Attarzadeh [1,2] and C. V. Ramana [1,3,*]

[1] Center for Advanced Materials Research, University of Texas at El Paso, 500 W. Univ. Ave., El Paso, TX 79968, USA; nattarzadeh@utep.edu
[2] Environmental Science and Engineering, University of Texas at El Paso, 500 W. Univ. Ave., El Paso, TX 79968, USA
[3] Department of Mechanical Engineering, University of Texas at El Paso, 500 W. Univ. Ave., El Paso, TX 79968, USA
* Correspondence: rvchintalapalle@utep.edu

Abstract: A plasma electrolytic oxidation (PEO) is an electrochemical and eco-friendly process where the surface features of the metal substrate are changed remarkably by electrochemical reactions accompanied by plasma micro-discharges. A stiff, adhesive, and conformal oxide layer on the Zr and Zr-alloy substrates can be formed by applying the PEO process. The review describes recent progress on various applications and functionality of PEO coatings in light of increasing industrial, medical, and optoelectronic demands for the production of advanced coatings. Besides, it explains how the PEO coating can address concerns about employing protective and long-lasting coatings with a remarkable biocompatibility and a broad excitation and absorption range of photoluminescence. A general overview of the process parameters of coatings is provided, accompanied by some information related to the biological conditions, under which, coatings are expected to function. The focus is to explain how the biocompatibility of coatings can be improved by tailoring the coating process. After that, corrosion and wear performance of PEO coatings are described in light of recognizing parameters that lead to the formation of coatings with outstanding performance in extreme loading conditions and corrosive environments. Finally, a future outlook and suggested research areas are outlined. The emerging applications derived from paramount features of the coating are considered in light of practical properties of coatings in areas including biocompatibility and bioactivity, corrosion and wear protection, and photoluminescence of coatings

Keywords: plasma electrolytic oxidation; biocompatibility; corrosion protection; wear resistance; photoluminescence

Citation: Attarzadeh, N.; Ramana, C.V. Plasma Electrolytic Oxidation Ceramic Coatings on Zirconium (Zr) and Zr-Alloys: Part-II: Properties and Applications. *Coatings* **2021**, *11*, 620. https://doi.org/10.3390/coatings11060620

Academic Editor: Małgorzata Norek

Received: 13 May 2021
Accepted: 18 May 2021
Published: 22 May 2021

Publisher's Note: MDPI stays neutral with regard to jurisdictional claims in published maps and institutional affiliations.

Copyright: © 2021 by the authors. Licensee MDPI, Basel, Switzerland. This article is an open access article distributed under the terms and conditions of the Creative Commons Attribution (CC BY) license (https://creativecommons.org/licenses/by/4.0/).

1. Introduction

Plasma electrolytic oxidation (PEO) is a promising electrochemical technique to form ceramic coatings using plasma-assisted oxidation on a broad range of metals, such as Mg, Ti, and Zr [1–3]. The significant merit of this technique is to fabricate protective inorganic layers grown from the substrate with the participation of species from the electrolyte. Forming PEO layers on valve metals boosts surface performances in many applications by enhancing the structural reliability and extending the longevity of functional properties. During the PEO process, the plasma state following extreme voltage and temperature transmits to the surface of the anodic substrate in an alkaline electrolyte and produces inorganic layers with outstanding adhesive strength to the underlayer [4,5]. As asserted unanimously, the PEO technique forms an environmentally friendly, wet-based coating implementing alkaline electrolytes as the primary media and providing conditions for the incorporation of inorganic additives, including silicate, phosphate, aluminate, or organic additives, including sodium oxalate, glycerol to facilitate the physicochemical attraction during the process via electrochemical reactions [6].

We explained the paramount roles of the electrical, compositional variables in determining the characteristics of coatings formed during the PEO process. These variables, such as current mode and density, waveform, and duty cycle, would considerably influence the alteration of coatings' surface and composition features. Therefore, earlier studies have been reported to shed light on the effects of PEO parameters on microdischarge characteristics and obtained features of inorganic layers on the Zr and Zr-alloys. Tailoring process variables during the PEO results in improving the applicability and enhancing the functionality of coatings in various aspects. For instance, the excellent corrosion properties of Zr-alloys were attained via controlling both electrical variables and electrolyte composition, which could, in turn, influence the spatial, lifetime, and electrical characteristics of the PEO coatings [7–13]. To the best of our knowledge, no comprehensive review has discussed the applications of the PEO method to prevent degradation, corrosion, and wear, and to improve the biocompatibility of Zr and Zr-alloy extensively. The recent advances in biological responses of Zr implants using the PEO coatings have been reviewed by Fattah-alhosseini's group recently [3]. Furthermore, we presented a detailed review of PEO coatings in Part-I of this issue, where the attention given was on deriving a deeper fundamental understanding of the PEO growth mechanisms and the effect of process parameters on transient discharge behavior at breakdown, initiation, and growth of the oxide layer, and on the incorporation of species from electrolyte. Part-I highlights the fundamental microstructural aspects associated with structural defects, phase transformation, and the role of additives. However, a detailed report on the properties and application of PEO coatings is still missing.

Considering the scarcity of reviews on the PEO coating of Zr and Zr-alloys as a specific group, a comprehensive review is strongly required to provide the integrated perspective on the applicability and performance of PEO coatings respective to current demands in industries, medication, and science. Accordingly, the present review provides the functional features of PEO to approach studying the applicability of coatings with special emphasis on imperative phenomena and applications of interest, such as biocompatibility in medical implantation, extensive endurance of coatings under applied loadings, outstanding protection of underlayer in exposure to corrosive environments, and a broad photoluminescence range. We describe the general picture of structural reliabilities in correspondence with tribological and electrochemical reactions, and the emerging frame of functional properties in a broad range of emitted light is also provided. Finally, future outlook and prospective progress in synergy with enhancing both pre- and post-treatments are given to provide advanced functionality of PEO coatings are outlined.

2. Biocompatibility of PEO Coatings

There is a significant desire to substitute Ti with Zr for surgical implant materials because of lower Young's modulus (92 GPa) compared to titanium (110 GPa), and hence, lower risk for implantation failure due to stress shielding [14,15]. The biocompatibility of Zr is related to the formation of an inherent oxide layer of zirconia, which is bio-inert and restricts its integration with bone tissue during implantation [16]. Therefore, direct bonding with bone and stimulating new bone formation on the Zr surface is not feasible. Besides, the inherent oxide layer with 2 to 5 nm thickness is prone to rupture, erosion, and wear once Zr is employed for high load-bearing implant applications [17,18]. To address these drawbacks, several surface modifications are implemented to boost the bioactivity and corrosion resistance of Zr without compromising its biocompatibility [18–22]. Out of numerous techniques, the PEO process has demonstrated a promising technique to develop firmly adherent, porous, crystalline, relatively rough, and thick oxide coatings on Zr in environmentally friendly alkaline based electrolytes.

Generally, an implant needs to demonstrate outstanding surface features, such as superior bioactivity and biocompatibility. It is expected that implants applicable to dental and orthopedic surfaces demonstrate the capability to facilitate new bone formation and bond steadily and firmly with bone following implantation [23]. The procedure to examine the apatite-forming ability of PEO coatings and Zr substrate is that they are immersed in simulated body fluid (SBF) medium for 14 days, as shown in Figure 1a. After that, the surface can be characterized using techniques, such as scanning electron microscopy-energy dispersive spectroscopy (SEM-EDS), for studying mineralization of the formed apatite layer on the coating surface and checking the elemental composition of the layer. The SEM image illustrated in Figure 1a showed formation of apatite compounds after immersion in SBF for 14 days. Precipitation of compounds composed of Ca and P over the surface was considered the indication of apatite formation. It was evident that sphere-shaped deposits covered the entire surface of all PEO coatings. However, the precipitated layer on the bare Zr surface was not uniform. Elemental analysis showed that the Ca/P ratios of precipitates on PEO coatings were in the range of 1.53 to 1.65, which is close to the stochiometric Ca/P ratio of 1.67 for hydroxyapatite [24].

Yan et al., took advantage of a post-treatment using ultraviolet (UV) irradiation to change the surface chemistry of the PEO coating. It turned out that strong UV irradiation caused the formation of more basic Zr-OH groups on the surface, and, therefore, the hydrophilic nature of the surface improved significantly [25]. Han et al., produced PEO coating containing Ca and P and studied the in-vitro bioactivity, and dominant osteoblast response of the film as a function of the applied voltage ranging from 400 to 500 V [26]. The PEO coating with a predominantly t-ZrO_2 phase could facilitate the accumulation of a more significant quantity of CaO and P compounds at higher applied voltage, while the apatite-forming ability of coatings was also boosted. A facile method to examine bioactivity and biocompatibility of PEO coatings was to study the proliferation and growth feasibility of the attached osteoblast cells on the surface in an alkaline phosphate environment [26]. It was deduced that doping ZrO_2 with Ca^{2+} ions could improve the accumulation of hydroxyapatite (HA) on the ZrO_2 surface. Yan et al., also worked on enhancing the apatite-forming ability by activating the PEO surface using acidic or alkaline solutions. Significant differences in formation of HA on the surface were noticed between the etched PEO surface and intact surface, where HA formed after a short time on the treated surface [27]. Zhang et al., also introduced HA nanorods into the PEO coating via a process in which HA nanorods were prepared via hydrothermal treatment over the PEO coating [28]. The produced calcium partially stabilized zirconia (Ca-PSZ) coating became covered with HA nanorods and showed greater hydrophilicity and apatite-forming ability. That coating was usually prepared in multistep methods or did not contain a crystalline HA layer on the substrates [27,28]. Cengiz et al., prepared a crystalline HA layer using a single-step PEO process on zirconium with enhanced bioactivity [29]. Composition characterization of PEO coatings showed that hydroxyapatite (HA) formed directly in the coating along with calcium zirconium oxide ($Ca_{0.134}Zr_{0.86}O_{1.86}$) Ca-PSZ and m-$ZrO_2$. The formation mechanism of the PEO coating with three layers is displayed in Figure 1b. It was found that Ca and P significantly accumulated in the outer region of the coating, while the inner region was rich in Zr and O. The reaction between pre-formed ZrO_2 and Ca^{2+} ions from the electrolyte resulted in formation of the Ca–Zr–O phase in the transition region. After that, high energy sparking regimes during the PEO process could induce HA layer formation, which could be initiated by complex reactions between Ca^{2+} and PO_4^{3-} ions on the anode surface. Thus, coatings with higher wettability could form during the PEO process. Figure 2a,b compares the wettability between the Zr substrate and the modified PEO coating. The formation of hydrophilic HA coating enhanced the wettability and reduced contact angle values in comparison with a bare Zr surface [29].

The crystallinity and morphology of HA phases could play a vital role in cell growth and cell activity [30,31]. Non-crystalline calcium phosphates were not favorable as they might be dissolved in human body fluids, implying instability for implantation [30,32]. Fu et al., studied 3T3 cell proliferation as a marker to test the biocompatibility of PEO coatings. They were able to measure 3T3 proliferation by MTS assay for PEO coatings formed in electrolytes with different KOH concentrations. The cell proliferation ability was enhanced significantly after seeding 3T3 cell on the PEO surface, however, the growth rate of 3T3 cells were similar regardless of their formation electrolytes [33].

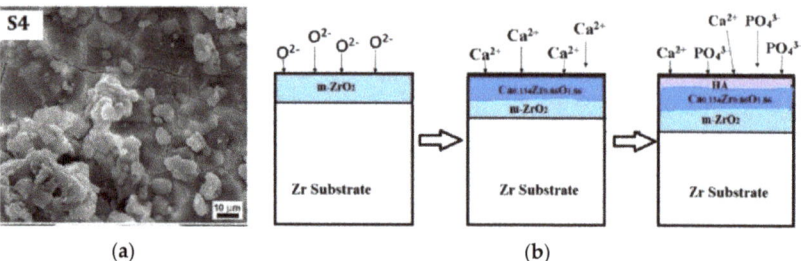

Figure 1. (a) SEM micrographs of the surface of the PEO coating immersed in SBF for 14 days. Formation of apatite particles is the indication of biocompatibility [24] (Reproduced with permission number: 5043530687884, Elsevier); (b) schematic representation of the PEO formation containing HA top layer [29] (Reproduced with permission number: 5061321012937, Elsevier).

In contrast to the many efforts to increase biocompatibility of PEO coatings on Zr, one of the other aims for surface improvements was to decrease bacterial activity or colonization in the vicinity of implants. Antibacterial activity of PEO coatings on Ti was enhanced by adding silver (Ag), thereby incorporating as an antibacterial agent [34,35]. Fidan et al., produced PEO coatings with silver acetate additives to halt bacterial activities of methicillin-resistant *Staphylococcus aureus* (MRSA) on the coatings [36]. Nanoparticles of Ag at concentration below toxicity levels could penetrate environments containing bacteria and expedite the release of Ag^+ ions, due to the greater specific surface. As shown in Figure 2c, Ag^+ ions demonstrated outstanding antibacterial activity against MRSA as one of the dominant reasons for infection after implantation. Durdu and his co-worker also worked on antibacterial properties of Ag-based bioceramic coating, where they deposited a thin layer of Ag hydrothermally on PEO coated Zr. They found that Ag distributed uniformly on the surface and that the final surface showed hydrophobic nature, in which the bacterial adhesion of the final surface reduced significantly compared to the condition without the Ag thin layer [37].

Aktuğ et al., produced PEO coatings composed of c-ZrO_2, perovskite-$CaZrO_3$, and hydroxyapatite (HA) in a single step for 5, 10, and 15 min [38]. They aimed to study the effect of PEO treatment time on the fabrication of a bioceramic-based hydroxyapatite coating. They studied the adhesion of Gram-positive and Gram-negative bacteria, including *E. coli, P. aureginosa, P. putida, B. subtilis, S. aureus*, and *E. faecalis* on the PEO coating. Hemolysis and MTT assays were used to examine the biocompatibility of coatings. Besides, the bioactivity was also tested using an in vitro immersion test for up to 28 days. Authors found that PEO surfaces demonstrated particular bioselectivity and modified cell–surface interactions. The microbial adhesion decreased with treatment time due to alteration of surface properties in thicker coatings. The difference in cell adhesion activity for Gram-negative and Gram-positive bacteria could be related to difference in cell surface hydrophilicity obtained by the higher amount of HA in prolonged processes [38]. Later, the same research group continued working on the production of PEO coatings containing HA composition to accelerate their interaction and bioactivity once they would be implanted [39].

Cengiz et al., studied the relation between microdischarge types and bioactivity of the PEO coatings for different treatment times ranging from 2.5 to 30 min [40]. The surface characterization revealed that the tetragonal phase formed as the dominant phase during all process times in electrolyte-containing $Ca(CH_3COO)_2 \cdot H_2O$ and $C_3H_7Na_2O_6P \cdot 5H_2O$. It was found that the amorphous HA formed during the PEO process on the coating surface facilitated the formation of HA crystals once coatings were immersed in SBF solution [40]. The combination of the PEO process and electrophoretic deposition (EPD) was also implemented to form bioactive ZrO$_2$/HA composite film [41]. Negatively charged HA nanoparticles were added to phosphate electrolyte to participate in a single-step PEO-EPD process for 2 to 6 min. The coating characterization showed that the uniform and dense ZrO$_2$/HA coating consisted of c-ZrO$_2$ and m-ZrO$_2$ nanocrystalline phases. HA nanoparticles were attracted to the discharge channels due to their charge and entrapped in channels because of the EPD process, while coatings supported greater wettability. After immersion into SBF for 8 days, an apatite layer covered throughout the surface of ZrO$_2$/HA coating, indicating considerable modification of in-vitro bioactivity. Besides, it was revealed that the human osteosarcoma could attach, adhere, and propagate considerably on the surface of the ZrO$_2$–HA layer, implying the possibility of its implementation as orthopedic implant material [41].

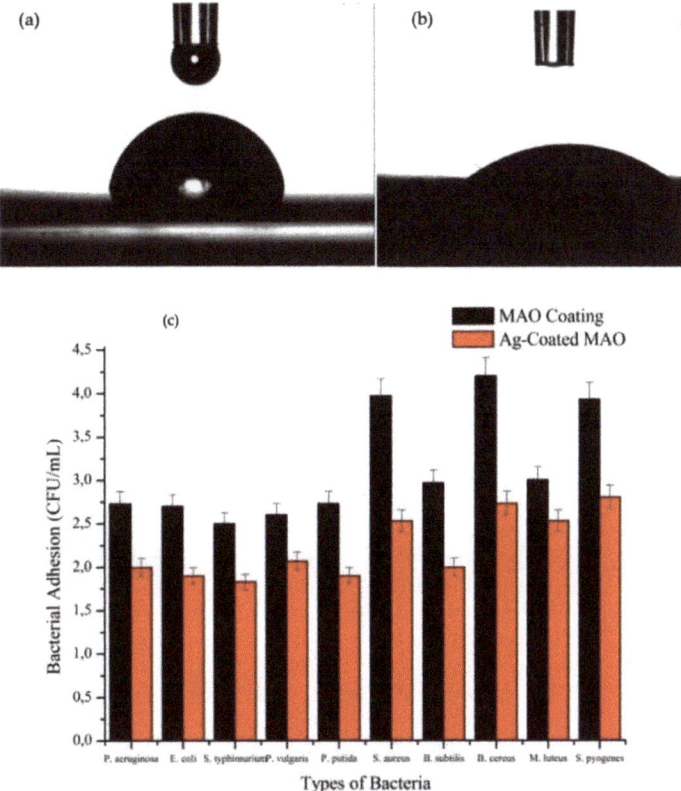

Figure 2. (**a**,**b**) The contact angle measurements: bare substrate and the PEO coating are displayed, respectively [29] (Reproduced with permission number: 5061321012937, Elsevier); (**c**) the plot for illustration of Gram-positive and Gram-negative bacteria's adhesion on the surface of PEO and Ag-based PEO coatings from [37] (Reproduced with permission number: 5061400310129, Elsevier).

Durdu et al., worked on thermal deposition of the Zn thin film on the PEO coating produced in electrolyte containing calcium acetate and β-CaGP salt [42]. The characterization of phase composition of coatings revealed that c-ZrO$_2$, CaZrO$_3$, and HA formed during the PEO process. The deposition of a uniform Zn layer on the entire surface caused super hydrophilicity compared to the undeposited surface. After immersion in SBF for 10 days, a compact and uniform bone-like apatite formed on the Zn-bioceramic surface. Microbial adherence tests for Gram-negative and Gram-positive bacteria exhibited that the antibacterial activity of the Zn-deposited surface was significantly strong against both types of bacteria, indicating enhanced bioactivity and antibacterial properties for the Zn-deposited PEO coating [42]. Zhang et al., fabricated HA nanorod-patterned zirconia coatings via a hybrid approach [43]. They found that the HA nanorod-patterned coating increased the feasibility of protein absorption and significantly modified the adhesive and proliferative properties of fibroblasts in comparison to untreated coating, suggesting a potential application for percutaneous implants to facilitate the connection to the skin [43]. Recently, Cu was also deposited onto the PEO surfaces without inducing any alteration in surface morphology by using thermal evaporation [44]. Surface characterization revealed the formation of meta-stable calcium zirconium oxide, cubic zirconia, and perovskite calcium zirconate phase. After completing the coating process, coatings with and without Cu top-layer were immersed in simulated body fluid to evaluate their in-vitro bioactivity and antimicrobial properties for Gram-positive and Gram-negative bacteria. Deposition of Cu hydrothermally on the PEO surface enhanced hydrophobic features, where both types of bacteria had a shorter microbial adhesion time to the surface of Cu-deposited PEO coating, thereby suggesting the biomedical application of the surface-modified PEO coating [44].

Very recently, the effect of pore size was studied in PEO coatings with respect to bioactivity and cell interaction properties [45]. Cengiz et al., prepared PEO coatings with various surface morphologies in terms of pore size and surface roughness to investigate in vitro bioactivity and cell interaction features after prolonged immersion in SBF. The in vitro bioactivity evaluations revealed the pore size of the coating significantly affected the growth rate of the HA layer on the coating surface. Remarkably, increasing the pore size promoted HA layer formation in SBF solution, in which higher roughness created by the PEO process increased the chance of HA nucleation, due to boosting the ionic activity of the coating surface [45].

3. Corrosion Resistance Properties

Nuclear grade Zr-alloys like Zircaloy-4 alloy are subjected to neutron irradiation as well as severe conditions of high temperature and high-pressure flowing water [46]. The harsh, aggressive environment exacerbates the corrosion and hydrogen absorption, thus causing fast deterioration of the mechanical strength of structural components [47,48]. Therefore, it is necessary for the nuclear industry to seek methods of curbing the corrosion rate of Zr alloys. On the other hand, Zr and its alloys have recently found a promising application as medical implants. The tendency of implant materials to be subjected to electrochemical reactions within body fluids is crucial to understanding their stability in the human body. Several techniques can be implemented to mimic the situation in the body and study the behavior of coated or noncoated Zr surfaces. Open circuit potential (OCP), Tefel extrapolation, potentiodynamic polarization (PDP), and electrochemical impedance spectroscopy (EIS) are the typical tests carried out on the PEO coated Zr samples to examine their corrosion performance in simulated body fluid (SBF) or other corrosive environments.

The PEO process is identified as a relatively fast, facile, and uncomplicated technique to produce functional ceramic coatings for corrosion-resistant applications on valve metals such as Ti and Zr. It is evident that the corrosion protection of the bare Zr substrate is associated with forming a native oxide layer on its surface. However, this oxide layer is not thick enough (i.e., in the order of a few tens of nanometer) and can be disappeared or degraded in the early stage of load bearing in prolonged applications like medical implants. This ends up raising the corrosion rate, thus reducing the efficacy and service time of the Zr implant. Although PEO coating is an effective method for lowering the corrosion rate due to blocking the charge-transfer mobility across the substrate–environment interface, the PEO coating suffers from its inherited porous nature, where deep pores provide an open path for penetration of aggressive ions and reduction in the overall protection efficiency of coating [9–13,24,49–52]. The PEO coating can form in various electrolytes, such as silicate, aluminate, phosphate, and so on, but all suffer from pancake-like surface features, pores, and cracks [53–55]. The rough and porous surface morphology did not disappear even for procedures without creating pancake-like structures [13,56]. Xue et al., explored the corrosion performance of PEO coating formed in silicate electrolyte using potentiodynamic polarization (PDP) tests [57]. They found that corrosion potentials shifted to more noble potentials and that the corrosion current density dropped several orders of magnitude due to the PEO coating. They claimed thickening the PEO coating impacted the corrosion rate, due to the higher number of cracks [57].

Sandhyarani et al., conducted a series of corrosion experiments to examine the corrosion performance of PEO coatings produced in the phosphate electrolyte with different treatment times. They found that the recorded OCPs of PEO coatings shifted to more noble potentials with a remarkable decrease in i_{corr} compared to that of the untreated Zr sample [12]. No pitting corrosion was observed for the PEO coating produced below 6 min treatment time. However, the PEO coating produced with the longer treatment time (8 min) suffered significant pitting corrosion, which was attributed to numerous cracks on the coated surface. The optimum corrosion performance in SBF environment was recorded for the Zr sample coated up to 6 min, while precipitates were mainly composed of Ca–P rich compounds [12].

The corrosion performance of the PEO coating produced on Zircaloy-4 in the silicate electrolyte using AC regime was tested in nitric acid solution [11]. The authors studied the influence of the frequency of the PEO process on the corrosion resistance of the formed coatings. Electrochemical studies revealed that the performance of coatings produced at both 50 and 100 Hz for 5 and 30 min were similar. No significant difference in the corrosion resistance was found for coatings produced at different frequencies, thus confirming the effect of similar coating morphologies. The Tafel test showed the current density declined remarkably for the PEO coating compared to the bare zirconium, and no noticeable difference was found between coatings with different thicknesses. The authors, therefore, deduced that the main protection was supported by the barrier layer at the base of the coatings [11].

The corrosion performance of PEO coatings produced in $Na_2SiO_3 \cdot 9H_2O$ and KOH solution with 1, 3, 10, and 30 min treatment times was tested in 0.5 M NaCl solution. Among PEO coatings, the coating with 10 min treatment time demonstrated the best corrosion protection properties. Increased porosity was the determining factor in decreasing the corrosion resistance of the PEO coating with 30 min treatment time [51].

Sandhyarani et al., also evaluated the effect of electrolyte composition on the corrosion performance of PEO coatings. They recorded OCP of PEO coatings formed in phosphate and aluminate electrolytes for 4 h and showed OCP values of all PEO coatings tended to more noble potentials, compared to Zr substrate, indicating higher thermodynamic stability of oxide films. At the beginning of immersion, OCP values for PEO coatings showed less noble potential and declined slightly. This instability at the early time of immersion could be associated with an increased activity because of penetration of corrosive solution from surface defects and pores to the metal–oxide interface. The oxide film formed in phosphate and KOH electrolyte showed the highest noble potential values over the entire period of immersion time, which can be related to the modified coating morphology of being pore-free and much smoother than other coatings. However, the film formed in silicate and KOH electrolyte showed higher apatite-forming ability because of superior wettability.

The evaluation of corrosion performance of the PEO composite coatings reiterated the importance of adding nanoparticles, in which the PEO composite coating produced with Ce_2O_3 addition could reduce the corrosion current density about 10^3 times compared to the simple PEO coating [9]. Figure 3a shows the polarization curves for the Zr substrate, the simple PEO, and the PEO composite coating with Al_2O_3, CeO_2, or ZrO_2 nanoparticles, termed as PA, PC, and PZ, respectively. Besides, the composite coating of ZrO_2–SiC produced in phosphate electrolyte exhibited superior corrosion resistance compared to coatings produced in aluminate and silicate electrolytes, due to the modified surface morphology of the coating [58].

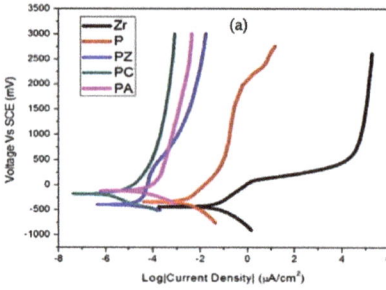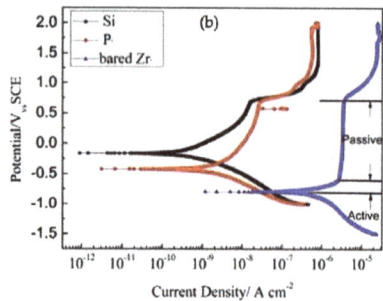

Figure 3. (a) PDP plot of Zr substrate, simple PEO coating (P), composite coating of PEO and each of Al_2O_3 (PA), Ce_2O_3 (PC), and ZrO_2 (PZ) [9] (Reproduced with permission number: 5043540009057, Elsevier); (b) PDP plot of bare zirconium, phosphate (P) and silicate (Si) PEO coatings [59] (Reproduced with permission number: 5043540272527, Elsevier).

Wang et al., conducted a comparative study on two types of PEO coatings produced on Zircaloy-4 alloy in silicate and phosphate electrolyte. The authors worked on the PDP behaviors and electrochemical impedance spectroscopy (EIS) measurements of PEO coatings immersed in LiOH solution for different time intervals to study the prolonged corrosion performance [59]. The surface morphology of the two coatings showed distinct differences, where the PEO coating formed in silicate electrolyte was more compact, and the population of pores and cracks was fewer than the PEO coating formed in phosphate electrolyte. The PEO coating formed in phosphate electrolyte suffered from localized peeling regions caused by stress mismatching of the considerable volume change owing to the transition from tetragonal to monoclinic phases. Figure 3b illustrates the PDP analysis of PEO coatings and the bare Zr substrate, which displayed no passive region for the PEO coatings as a remarkable difference with the bare sample. The corrosion current density is three and four orders of magnitude lower than that of the bare electrode for PEO coatings formed in phosphate and silicate electrolyte, respectively.

EIS measurements were performed for both coatings immersed in LiOH solution, and various equivalent electrical circuits were suggested to simulate results. Figure 4a,b shows the Nyquist plots of EIS results for silicate and phosphate coatings, and Figure 4c,d shows the suggested electrical circuits for prolonged immersions. It was evident that the silicate coating with modified surface morphology could endure the aggressive condition greater and protect zirconium more significantly in the long-term immersion process.

Li et al., mixed graphene oxide (GO) particles into silicate and a mixture of phosphate and silicate electrolytes to produce the PEO composite coatings of zirconia with graphene on N36 Zr alloy [8]. They studied corrosion and fretting behavior of the composite coating and investigated the influence of GO addition on improving the corrosion resistance. The authors reported the positive role of GO particles in enhancing the anti-corrosion behavior of PEO coating, in which the coating formed with 0.1 g/L GO at silicate electrolyte exhibited the best corrosion performance. The GO particles developed a good barrier against penetration of aggressive ions and blocked access to the substrate [8]. Very recently, the same authors compared the corrosion performance of PEO composite coating with each of Al_2O_3, MoS_2, CeO_2, and graphene oxide (GO) on Zr alloy [7]. The positive effect of adding nanoparticles was to lower the corrosion current density with more noble corrosion potential due to impeding aggressive ions before they reach the substrate.

Wu et al., combined the PEO process with pulsed laser deposition (PLD) to enhance the corrosion protection of composite film formed on Zr-4 alloy [60]. PLD was used to form a $Cr/CrN/Cr_2O_3$ film covering the PEO coating to boost the corrosion resistance of the film on Zr alloy. The EIS study confirmed improving the corrosion resistance via enlarging the polarization resistance, suggesting contribution of dense $Cr/CrN/Cr_2O_3$ film on the corrosion performance of ZrO_2 film though blocking ion attacks more effectively.

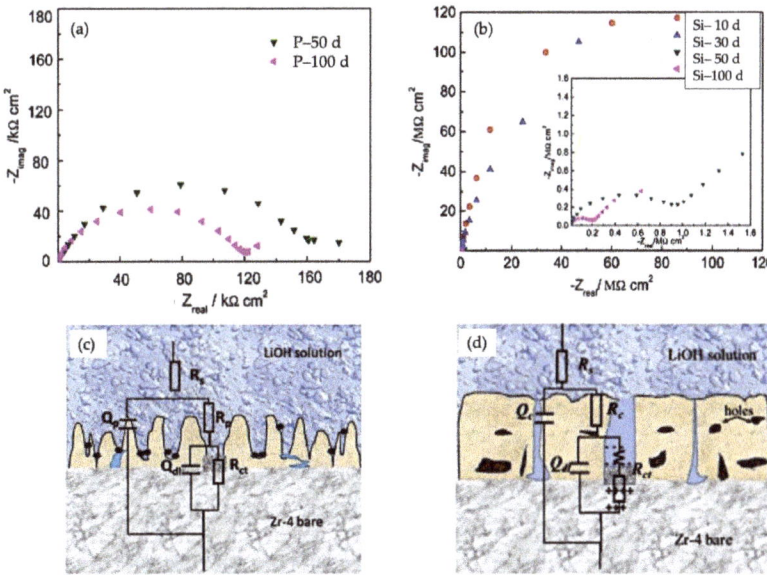

Figure 4. (**a**,**b**) Nyquist plots of EIS results of phosphate and silicate PEO coating for prolonged immersions, respectively [59]; (**c**,**d**) Equivalent circuit and physical model for phosphate and silicate PEO coatings after 50–100 days immersion, respectively [59] (Reproduced with permission number: 5043540272527, Elsevier).

4. Wear and Tribology of the PEO Coatings

Zr alloys demonstrated low neutron absorption coefficients and high yield strength at higher temperatures, which led to their selection for nuclear applications, such as fuel tubes and coolant channel materials in water-cooled power reactors [10,46,61,62]. Furthermore, Zr alloys suffer from low wear resistance in comparison with other nuclear materials, such as stainless steels and nickel alloys [63], while the fretting wear on tubes is an important concern in pressurized water reactors [63,64]. Besides, Zr alloys also demonstrated potential for biomedical applications, and thus, improving wear resistance can allow Zr-based implants to replace Ti-based implants. The dry sliding wear performance of the PEO coating can be related to main properties, including microstructure, composition, thickness of coatings, and the sliding procedures, which are related to the applied load and sliding time. Figure 5a shows the wear scar of the PEO coating on Zircaloy-2 in silicate electrolyte for 20 min. Due to the sliding wear test, it was evident that the coating was completely peeled off after 600 s, and the ploughing effects were visible on the substrate. As shown in Figure 5b,c, the PEO coating formed in aluminate electrolyte for 10 min is not worn out fully even after 30 min of sliding wear test [53]. The weak performance of the PEO coating formed in silicate electrolyte could be attributed to numerous cracks and pores in the coating, while the PEO coating formed in aluminate electrolyte showed greater wear performance due to the coating's compactness and higher t-ZrO_2 content.

Cheng et al., performed dry sliding wear tests using an applied load of 10 N for coatings produced for 5, 10, and 30 min in dilute electrolyte of aluminate and KOH [54]. The friction coefficient of the PEO coating formed for 5 min ranged from ~0.13 to ~0.17 during 240 s of the sliding test; a continuous rise was then observed in the friction coefficient after the initial low values, which could indicate substrate removal. During the initial stage of testing, the very low friction coefficient was driven by the modified morphology of the coating. It was found that the PEO coating formed for 10 min showed weaker wear protection under the same load. The large gap (~20 μm) between the inner layer and the outer layer caused the weak tribological behavior. The transition point (end of coating life) for the coating formed for 10 min occurred much faster than the coating formed for 5 min. Finally, the friction coefficient of the coating formed for 30 min rose significantly from ~0.4 to ~0.67 during 344 s as the initial stage of testing. After that, the coefficient remained constant until ~1252 s, and then abruptly declined to the lowest value of ~0.47 at 1394 s. Figure 5d displays the profile along the cross-sections of the wear scars after the dry sliding tests for coatings produced with different PEO treatment times. The weakest wear performance with the deepest scar was observed for the PEO coating produced for 10 min due to the early collapse of the compact outer layer, and thus exposing the thin inner layer. The greatest wear performance was recorded for the coating formed for 30 min with a wear depth of ~20 μm.

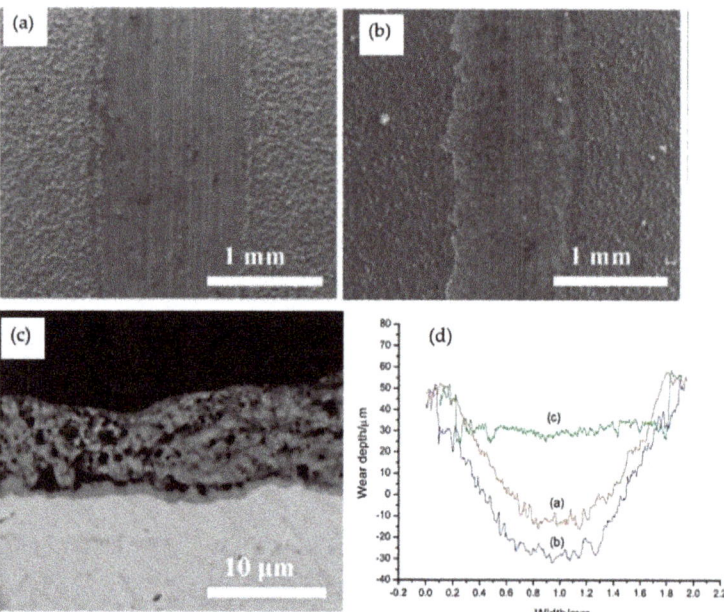

Figure 5. (a,b) The sliding impact on surface morphologies for the silicate and aluminate PEO coating, respectively; (c) cross-section image of the sliding scar at the center of the PEO coating formed in aluminate electrolyte for 600 s [53] (Reproduced with permission number: 5043480179625, Elsevier); (d) wear scars profile after dry sliding under 10 N load on PEO coatings formed in electrolyte mixture of aluminate and KOH for various treatment times: a. 5 min, 30 min sliding time; b. 10 min, 923 s sliding time; c. 30 min, 30 min sliding time [54] (Reproduced with permission number: 5043510569542, Elsevier).

The surface cross-section image of the wear scar remained after dry sliding wear test for the PEO coating produced in aluminate electrolyte for 30 min were shown in Figure 6a. It was evident that the whole outer layer was removed, while the inner layer was intact. The outstanding performance of the coating stemmed from the mechanical strength of the outer layer and the great adhesion of the inner layer to the substrate [54,65]. Figure 6b shows the impact of the sliding during wear testing from the central part of the wear scar. Due to extended sliding, an iron oxide layer remained on the wear track. The patches with the transfer layer protruded above the surrounding inner layer and supported the contact points with the steel ball. The remnant debris from the previous round of sliding enhanced the wear resistance during the sliding process and acted as lubrication [65]. The sliding wear tests were also performed on the coatings formed in concentrated aluminate electrolyte [54]. Furthermore, the PEO coating formed for 10 min in concentrated aluminate electrolyte with the thickness of ~64.4 µm was tested under a high load of 30 N and compared with the coating formed in dilute aluminate electrolyte for 30 min with a thickness of ~75.8 µm. Figure 6c,d demonstrate the cross-sectional profiles of wear scars due to the dry sliding test for 30 min for comparing PEO coatings formed in concentrated and dilute aluminate electrolyte and for comparing PEO coatings formed in aluminate and phosphate electrolyte [54]. The low friction coefficient at early times could be related to lack of formation of the transfer layer on the coating surface; however, developing the transfer layer due to frictional heating and mechanical stresses caused the friction coefficient to rise [66]. Cheng et al., finally concluded that the PEO coating formed for 10 min demonstrated greater wear protections, while the wear depth was ~20 µm after 30 min of dry sliding with the applied load of 30 N.

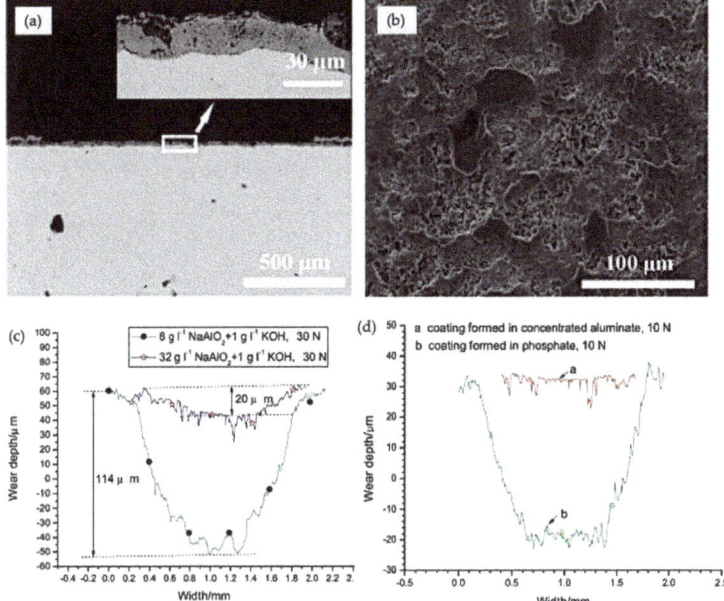

Figure 6. (a) The cross-section image of sliding bar impact on the coating surface formed in aluminate electrolyte [54]; (b) the central part of the wear scar on the PEO coating formed in aluminate electrolyte [54]; (c) the wear profile of the sliding test for PEO coatings produced in dilute and concentrated aluminate electrolyte under 30 N applied load [54]; (d) the wear profile of the sliding test for the PEO coating formed in concentrated aluminate electrolyte (red color) and the PEO coating formed in phosphate electrolyte (green color) under 10 N applied load [54] (Reproduced with permission number: 5046821419713, Elsevier).

The presence of nanoplate-like α-Al_2O_3 out-layered Al_2O_3–ZrO_2 composite PEO coating enhanced the wear protection of PEO coating significantly [67,68], highlighting its potential applicability for articular head replacement. It was found that the presence of tetragonal zirconia could enhance the protection performance of the coating produced in aluminate electrolyte, in which the applied load could expedite the transformation of zirconia from tetragonal to monoclinic phases. The volume expansion caused by the transformation induced dilatational and shear stresses, preventing the progression of cracks, and thus improving the wear performance [68].

Wei et al., worked on the wear mechanism of PEO coating produced in phosphate electrolyte [69]. They found that the lowest wear rate of the PEO coating was 1/60 of bare

Zirlo alloy in similar conditions, and the wear mechanism of PEO coating was primarily the adhesive wear, where the substrate demonstrated the abrasive wear [69]. Adding graphene oxide (GO) particles in electrolyte also showed a positive effect on reducing friction and improved the fretting wear resistance [8]. The PEO coating with GO addition showed delamination, wear debris, and channel trace, demonstrating abrasive wear as the dominant wear mechanism. However, the PEO coating with GO did not show worn-out protrusion in the surface morphology. Recently, Li et al., conducted a comparative study on the influence of different nanoparticles, including Al_2O_3, MoS_2, CeO_2, and graphene oxide (GO), on fretting corrosion behavior [7]. Basically, the nanoparticles embedded in coatings acted as a lubrication phase to decrease the friction. The composite PEO coatings with MoS_2 and GO showed low coefficient of friction value in the initial stage of the wear test, after detaching the nanoparticles, the coefficient of friction rapidly increased. It was suggested that abrasive wear was the main wear mechanism for the PEO coating with Al_2O_3, while the wear mechanism for the PEO coating with either MoS_2 or CeO_2 was abrasive wear and adhesive wear. For the PEO coating with GO, the wear mechanism was abrasive wear. The PEO composite coating with GO showed that the wear volume was lowest among other composite coatings, but its tribocorrosion performance revealed the highest material loss due to the impact of wear during the corrosion test. The presence of GO particles could effectively reduce the friction and wear as GOs had small shear stress between the layers making the sliding process easier.

5. Photoluminescence Performance

ZrO_2 based materials have drawn considerable interest from many researchers in photonics, optoelectronics, and electronics due to their outstanding thermal, mechanical, electrical, and optical properties [70–73]. Optical transparency in the visible and near infrared region, high refractive index, low optical loss, wide energy bandgap (~5 eV), and low phonon frequency (~470 cm^{-1}) have demonstrated zirconia as a potential matrix host for fabrication of doped compounds using highly efficient luminescent materials ranging from rare-earth to transition metals [74–79]. Recently, researchers have intended to take advantage of PEO processing to entrap rare-earth oxide particles in the ZrO_2 matrix [74–81].

Figure 7a demonstrated diffuse reflection spectra of the PEO coating produced in citric acid electrolyte for a range of treatment times from 1 to 10 min [82]. A wide absorption band was detected spanning the range of 200 to 330 nm. The main absorption peak occurred at approximately 236 nm (~5.2 eV in photon energy) that could be related to photon excitation from valance band to conduction band [83,84]. In addition, the absorption peak at approximately 294 nm (~4.3 eV in photon energy) could be related to the interstitial Zr^{3+} ions [85]. Photoluminescence (PL) measurements displayed easily identifiable PL bands ranging from 300 to 600 nm at ambient temperature. The emission spectra divulged 4 peaks centered approximately at 418, 440, 464, and 495 nm. The prolonged PEO treatment time increased the PL intensity because the concentration of oxygen vacancies enhanced in m-ZrO_2 films, which was responsible for PL excitation bands. This behavior was also recorded for photocatalytic activity of zirconia films, in which prolonged PEO treatment time enhanced the activity as a result of entrapping more oxygen vacancies in the thicker film [82]. Stojadinović et al., worked on the possibility of fabricating doped ZrO_2 coatings with Ho^{3+} and Ho^{3+}/Yb^{3+} using the PEO process in phosphate electrolyte containing Ho_2O_3 or H_2O_3/Yb_2O_3 particles [78]. They investigated up- and down-conversion PL properties of the complex oxide coatings composed of ZrO_2 as a host and rare-earth oxide as components of the composite. Notably, rare-earth oxide particles featured zeta potentials in the negative range for alkaline environments, and thereby, the applied potential in the PEO process sucked Ho_2O_3 and Yb_2O_3 particles toward the anode and entrapped them inside microdischarge channels to form a composite of mixed oxides [78]. The PL emission spectra of PEO coatings evolved with different quantities of Ho_2O_3 particles and excited with 280 nm radiation, as illustrated in Figure 7b. Pure ZrO_2 coating could only display a wide PL band with a spectral maximum approximately at 490 nm in the visible region originating from optical transitions in PL centers derived from oxygen vacancies of ZrO$_2$ [82].

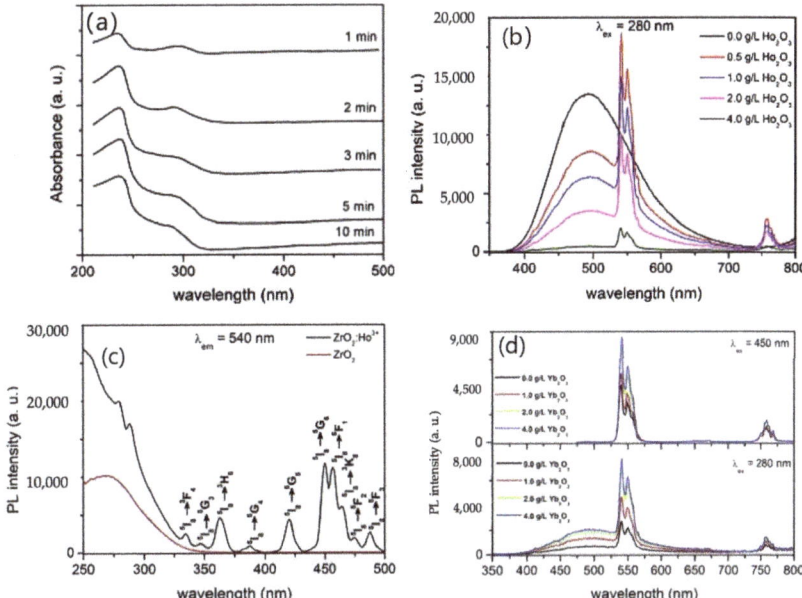

Figure 7. (**a**) Alteration of diffuse reflection spectra of zirconia films based on treatment time of the PEO process [82] (Reproduced with permission number: 5043541253698, Elsevier); (**b**) photoluminescence emission spectra excited with 280 nm radiation of PEO coating produced in phosphate electrolyte with addition of Ho$_2$O$_3$ powder in different concentrations for a treatment time of 10 min [78]; (**c**) photoluminescence excitation spectra of undoped and Ho-doped ZrO$_2$ coatings produced by PEO for a treatment time of 10 min in phosphate electrolyte with and without 0.5 g/L Ho$_2$O$_3$ addition [78]; (**d**) photoluminescence emission spectra excited with 280 and 450 nm radiation of PEO coatings produced in phosphate electrolyte with addition of 4.0 g/L Ho$_2$O$_3$ particles and Yb$_2$O$_3$ particles in different concentrations for a treatment time of 10 min [78] (Reproduced with permission number: 5043550012051, Elsevier).

Therefore, two overlapping regions could be found in down-conversion PL spectra of the composite coating excited by 280 nm including the broad PL bond from the matrix and several emission bands related to f–f transition of Ho^{3+} ions [78]. At around 540 and 550 nm, there were two predominant peaks relating to $^5F_4 \rightarrow {}^5I_8$ and $^5S_2 \rightarrow {}^5I_8$ transitions, respectively. Increasing treatment time of the PEO processing caused intensifying the wide emission bond of the PL intensity of the ZrO$_2$ matrix and enhancing the emission bonds of Ho^3. It was apparent that the PL intensity was greater for un-doped coatings, where higher concentration of Ho^{3+} ions participating in the ZrO$_2$ matrix caused reducing the PL intensity of the broad emission band, indicating the presence of a non-radiative energy transient between matrix and dopants [78]. The PL excitation spectra were recorded at 540 nm for both doped and un-doped zirconia coatings, as shown in Figure 7c. While the excitation PL spectrum of un-doped zirconia depicted a wide band centered at about 280 nm, the excitation spectrum of Ho-doped zirconia demonstrated an intensified wide band ranging from 250 to 325 nm and several sharp peaks between 325 and 500 nm. Electron transit between fully occupied 2p orbitals of O^{2-} ions of ZrO$_2$ and 4f orbitals of Ho^{3+} ions led to the formation of the early broadened band, while sharp peaks were related to direct excitation from $^5I_8$ ground state to the greater energy level of the 4f-manifold. Peaks and assigned electronic transition are clearly illustrated in Figure 7c. Therefore, the PL emission intensity of doped zirconia coatings with rare-earth ions could alter with changing the thickness of the PEO coatings and concentration of rare-earth ions participating in the composite coating [78]. The down-conversion PL emission spectra of PEO coatings formed in electrolyte containing H_2O_3 and different concentrations of Yb_2O_3 particles are also shown for excitements at 280 and 450 nm radiations in Figure 7d. The addition of Yb^{3+} ions into the composite of $ZrO_2:Ho^{3+}$ coatings did not alter the overall shapes of the PL spectra, however, the intensity of the wideband derived from the zirconia matrix and PL bands related to f–f transition of Ho^{3+} ions were enhanced by raising the incorporation of Yb^{3+} ions into the composite coatings. This behavior was an indication of the role of Yb^{3+} ions in transmitting energy to Ho^{3+} ions [78,86].

Ćirić and Stojadinović also produced the ZrO_2 coating with Pr^{3+} incorporated particles [76]. Accordingly, they found that increasing the treatment time of the PEO process also increased the incorporation of uniformly distributed Pr^{3+} in the composite coating, where its contribution concurrently stabilized the tetragonal zirconia phase. They categorized PL emission spectral features into two distinct regions, including a broad PL band centered at approximately 495 nm and secondly a region due to the 4f–4f transitions of Pr^{3+}, where the most intense one was from the $^1D_2 \rightarrow {}^3H_4$ transition [76]. The same authors reported fabrication of doped composite zirconia coatings containing Tm^{3+} and Yb^{3+} using the PEO process [75]. Tm and Yb particles promoted phase transformation from monoclinic to tetragonal phase, where the PL excitation spectra were identified by the broad band of ZrO_2 centered at approximately 280 nm and a peak at approximately 359 nm related to the $^3H_6 \rightarrow {}^1D_2$ transition of Tm^{3+}.

6. Concluding Remarks and Future Outlook

The PEO coatings have various applications upon the preparation process and zirconium substrate. Several methods have already been implemented to improve bioresponse and antibacterial activity on the Zr surface. Until now, the PEO process has demonstrated promising results in reducing the risks associated with using the PEO coated Zr. Particularly, medical implants demand significant surface biocompatibility, which bare zirconium cannot meet for certain criteria. The rough and porous surface of PEO coatings stimulates nucleation and growth of biocompatible hydroxyapatite compounds due to offering a larger available surface for ionic interactions. The formation of PEO coatings reduces bacterial adhesion, and the incorporation of antibacterial agents, such as silver ions, provides a greater antibacterial feature. Incorporation of Zn and Cu ions in the PEO coating enhances the biocompatibility of coatings, where a more uniform and compact hydroxyapatite forms on the surface of coatings after prolonged immersion in the simulated body fluid.

PEO coating can enhance the corrosion protection of zirconium when immersed in aggressive media. The coatings were able to block the penetration paths of aggressive ions and enhanced the endurance of the zirconium substrate. The PEO coating also reduced the coefficient of friction and offered greater protection against sliding wear, as the debris released from the coatings usually acted as lubrication. Besides, the PEO composite coating with rare-earth elements offered tuning photoluminescence features of the zirconia, where the addition of different rare earth oxides could tailor the excitation and emission range of composite zirconia oxide.

On account of reviewed studies, the following remarks can be suggested for future studies:

- Inducing new nanoparticles in electrolytes during the PEO process opens a new path for future research, where there is a lack of studies on biomedical applications of Zr and Zr-alloys, such as enhancing the coatings' bioresponse and efficient delivering drugs to the body tissues.
- Poor surface features such as numerous pores and cracks confine the protection mechanism to the barrier inhibition, in which the coating halts the penetration of aggressive ions. However, producing composite coatings with protective agents such as sacrificing anodic nanoparticles can endow active protection to the substrates.

- Wear behavior of coatings requires significant improvements. Besides, the wear mechanism of the coating has not been studied comprehensively. Using the PEO coated Zr-alloys in nuclear industries and applications demanding significant protection against fretting wear and corrosion requires more studies with simulated conditions.
- Tuning ZrO_2 bandgap for photocatalytic applications can draw attention to producing composite coatings with versatile applicability in splitting water and other compounds.

Author Contributions: C.V.R. conceived and supervised this work. N.A. collected all the data, obtained permission to use the data existing in the literature, and compiled the scientific validations. Both authors involved in preparing the manuscript. Both authors have read and agreed to the published version of the manuscript.

Funding: The authors also acknowledge, with pleasure, support from the National Science Foundation (NSF) with NSF-PREM grant #DMR-1827745.

Acknowledgments: NA acknowledges with pleasure the technical support and encouragement provided by the Center for Advanced Materials Research (CMR), UTEP. NA also acknowledges the Research Associate opportunity provided by CMR, UTEP.

Conflicts of Interest: The authors declare no conflict of interest.

References

1. Fattah-alhosseini, A.; Molaei, M.; Attarzadeh, N.; Babaei, K.; Attarzadeh, F. On the Enhanced Antibacterial Activity of Plasma Electrolytic Oxidation (PEO) Coatings That Incorporate Particles: A Review. *Ceram. Int.* **2020**, *46*, 20587–20607. [CrossRef]
2. Fattah-alhosseini, A.; Molaei, M.; Babaei, K. The Effects of Nano- and Micro-Particles on Properties of Plasma Electrolytic Oxidation (PEO) Coatings Applied on Titanium Substrates: A Review. *Surf. Interfaces* **2020**, *21*, 100659. [CrossRef]
3. Molaei, M.; Attarzadeh, N.; Fattah-alhosseini, A. Tailoring the Biological Response of Zirconium Implants Using Zirconia Bioceramic Coatings: A Systematic Review. *J. Trace Elem. Med. Biol.* **2021**, *66*, 126756. [CrossRef]
4. Bordbar-Khiabani, A.; Yarmand, B.; Sharifi-Asl, S.; Mozafari, M. Improved Corrosion Performance of Biodegradable Magnesium in Simulated Inflammatory Condition via Drug-Loaded Plasma Electrolytic Oxidation Coatings. *Mater. Chem. Phys.* **2020**, *239*, 122003. [CrossRef]
5. Wang, D.; Liu, X.; Wu, Y.; Han, H.; Yang, Z.; Su, Y.; Zhang, X.; Wu, G.; Shen, D. Evolution Process of the Plasma Electrolytic Oxidation (PEO) Coating Formed on Aluminum in an Alkaline Sodium Hexametaphosphate ($(NaPO_3)_6$) Electrolyte. *J. Alloy. Compd.* **2019**, *798*, 129–143. [CrossRef]
6. Sankara Narayanan, T.S.N.; Park, I.S.; Lee, M.H. Strategies to Improve the Corrosion Resistance of Microarc Oxidation (MAO) Coated Magnesium Alloys for Degradable Implants: Prospects and Challenges. *Prog. Mater. Sci.* **2014**, *60*, 1–71. [CrossRef]
7. Li, Z.; Cai, Z.; Cui, X.-J.; Liu, R.; Yang, Z.; Zhu, M. Influence of Nanoparticle Additions on Structure and Fretting Corrosion Behavior of Micro-Arc Oxidation Coatings on Zirconium Alloy. *Surf. Coat. Technol.* **2021**, *410*, 126949. [CrossRef]
8. Li, Z.; Cai, Z.; Ding, Y.; Cui, X.-J.; Yang, Z.; Zhu, M. Characterization of Graphene Oxide/ZrO_2 Composite Coatings Deposited on Zirconium Alloy by Micro-Arc Oxidation. *Appl. Surf. Sci.* **2020**, *506*, 144928. [CrossRef]
9. Arun, S.; Arunnellaiappan, T.; Rameshbabu, N. Fabrication of the Nanoparticle Incorporated PEO Coating on Commercially Pure Zirconium and Its Corrosion Resistance. *Surf. Coat. Technol.* **2016**, *305*, 264–273. [CrossRef]
10. Qin, W.; Nam, C.; Li, H.L.; Szpunar, J.A. Tetragonal Phase Stability in ZrO_2 Film Formed on Zirconium Alloys and Its Effects on Corrosion Resistance. *Acta Mater.* **2007**, *55*, 1695–1701. [CrossRef]
11. Cheng, Y.; Matykina, E.; Arrabal, R.; Skeldon, P.; Thompson, G.E. Plasma Electrolytic Oxidation and Corrosion Protection of Zircaloy-4. *Surf. Coat. Technol.* **2012**, *206*, 3230–3239. [CrossRef]
12. Sandhyarani, M.; Rameshbabu, N.; Venkateswarlu, K.; Sreekanth, D.; Subrahmanyam, C. Surface Morphology, Corrosion Resistance and in Vitro Bioactivity of P Containing ZrO_2 Films Formed on Zr by Plasma Electrolytic Oxidation. *J. Alloy. Compd.* **2013**, *553*, 324–332. [CrossRef]
13. Yan, Y.; Han, Y.; Li, D.; Huang, J.; Lian, Q. Effect of $NaAlO_2$ Concentrations on Microstructure and Corrosion Resistance of Al_2O_3/ZrO_2 Coatings Formed on Zirconium by Micro-Arc Oxidation. *Appl. Surf. Sci.* **2010**, *256*, 6359–6366. [CrossRef]
14. Gefen, A. Optimizing the Biomechanical Compatibility of Orthopedic Screws for Bone Fracture Fixation. *Med. Eng. Phys.* **2002**, *24*, 337–347. [CrossRef]
15. Thomsen, P.; Larsson, C.; Ericson, L.E.; Sennerby, L.; Lausmaa, J.; Kasemo, B. Structure of the Interface between Rabbit Cortical Bone and Implants of Gold, Zirconium and Titanium. *J. Mater. Sci. Mater. Med.* **1997**, *8*, 653–665. [CrossRef] [PubMed]
16. Liu, Y.-T.; Lee, T.-M.; Lui, T.-S. Enhanced Osteoblastic Cell Response on Zirconia by Bio-Inspired Surface Modification. *Colloids Surf. B Biointerfaces* **2013**, *106*, 37–45. [CrossRef] [PubMed]
17. Wang, L.; Hu, X.; Nie, X. Deposition and Properties of Zirconia Coatings on a Zirconium Alloy Produced by Pulsed DC Plasma Electrolytic Oxidation. *Surf. Coat. Technol.* **2013**, *221*, 150–157. [CrossRef]

18. Sanchez, A.G.; Schreiner, W.; Duffó, G.; Ceré, S. Surface Characterization of Anodized Zirconium for Biomedical Applications. *Appl. Surf. Sci.* **2011**, *257*, 6397–6405. [CrossRef]
19. Zhou, F.Y.; Wang, B.L.; Qiu, K.J.; Li, H.F.; Li, L.; Zheng, Y.F.; Han, Y. In Vitro Corrosion Behavior and Cellular Response of Thermally Oxidized Zr–3Sn Alloy. *Appl. Surf. Sci.* **2013**, *265*, 878–888. [CrossRef]
20. Brenier, R.; Mugnier, J.; Mirica, E. XPS Study of Amorphous Zirconium Oxide Films Prepared by Sol–Gel. *Appl. Surf. Sci.* **1999**, *143*, 85–91. [CrossRef]
21. Leushake, U.; Krell, T.; Schulz, U.; Peters, M.; Kaysser, W.A.; Rabin, B.H. Microstructure and Phase Stability of EB-PVD Alumina and Alumina/Zirconia for Thermal Barrier Coating Applications. *Surf. Coat. Technol.* **1997**, *94–95*, 131–136. [CrossRef]
22. Shanmugavelayutham, G.; Yano, S.; Kobayashi, A. Microstructural Characterization and Properties of ZrO_2/Al_2O_3 Thermal Barrier Coatings by Gas Tunnel-Type Plasma Spraying. *Vacuum* **2006**, *80*, 1336–1340. [CrossRef]
23. Zhang, Z.; Wang, K.; Bai, C.; Li, X.; Dang, X.; Zhang, C. The Influence of UV Irradiation on the Biological Properties of MAO-Formed ZrO_2. *Colloids Surf. B Biointerfaces* **2012**, *89*, 40–47. [CrossRef]
24. Sandhyarani, M.; Prasadrao, T.; Rameshbabu, N. Role of Electrolyte Composition on Structural, Morphological and in-Vitro Biological Properties of Plasma Electrolytic Oxidation Films Formed on Zirconium. *Appl. Surf. Sci.* **2014**, *317*, 198–209. [CrossRef]
25. Han, Y.; Yan, Y.; Lu, C. Ultraviolet-Enhanced Bioactivity of ZrO_2 Films Prepared by Micro-Arc Oxidation. *Thin Solid Film.* **2009**, *517*, 1577–1581. [CrossRef]
26. Han, Y.; Yan, Y.; Lu, C.; Zhang, Y.; Xu, K. Bioactivity and Osteoblast Response of the Micro-Arc Oxidized Zirconia Films. *J. Biomed. Mater. Res. Part A* **2009**, *88A*, 117–127. [CrossRef] [PubMed]
27. Yan, Y.; Han, Y.; Lu, C. The Effect of Chemical Treatment on Apatite-Forming Ability of the Macroporous Zirconia Films Formed by Micro-Arc Oxidation. *Appl. Surf. Sci.* **2008**, *254*, 4833–4839. [CrossRef]
28. Zhang, L.; Zhu, S.; Han, Y.; Xiao, C.; Tang, W. Formation and Bioactivity of HA Nanorods on Micro-Arc Oxidized Zirconium. *Mater. Sci. Eng. C* **2014**, *43*, 86–91. [CrossRef] [PubMed]
29. Cengiz, S.; Uzunoglu, A.; Stanciu, L.; Tarakci, M.; Gencer, Y. Direct Fabrication of Crystalline Hydroxyapatite Coating on Zirconium by Single-Step Plasma Electrolytic Oxidation Process. *Surf. Coat. Technol.* **2016**, *301*, 74–79. [CrossRef]
30. Surmenev, R.A.; Surmeneva, M.A.; Ivanova, A.A. Significance of Calcium Phosphate Coatings for the Enhancement of New Bone Osteogenesis—A Review. *Acta Biomater.* **2014**, *10*, 557–579. [CrossRef] [PubMed]
31. Viswanath, B.; Ravishankar, N. Controlled Synthesis of Plate-Shaped Hydroxyapatite and Implications for the Morphology of the Apatite Phase in Bone. *Biomaterials* **2008**, *29*, 4855–4863. [CrossRef]
32. Roy, M.; Bandyopadhyay, A.; Bose, S. Induction Plasma Sprayed Nano Hydroxyapatite Coatings on Titanium for Orthopaedic and Dental Implants. *Surf. Coat. Technol.* **2011**, *205*, 2785–2792. [CrossRef] [PubMed]
33. Lu, S.-F.; Lou, B.-S.; Yang, Y.-C.; Wu, P.-S.; Chung, R.-J.; Lee, J.-W. Effects of Duty Cycle and Electrolyte Concentration on the Microstructure and Biocompatibility of Plasma Electrolytic Oxidation Treatment on Zirconium Metal. *Thin Solid Film.* **2015**, *596*, 87–93. [CrossRef]
34. Bayati, M.R.; Aminzare, M.; Molaei, R.; Sadrnezhaad, S.K. Micro Arc Oxidation of Nano-Crystalline Ag-Doped TiO_2 Semiconductors. *Mater. Lett.* **2011**, *65*, 840–842. [CrossRef]
35. Teker, D.; Muhaffel, F.; Menekse, M.; Karaguler, N.G.; Baydogan, M.; Cimenoglu, H. Characteristics of Multi-Layer Coating Formed on Commercially Pure Titanium for Biomedical Applications. *Mater. Sci. Eng. C* **2015**, *48*, 579–585. [CrossRef]
36. Fidan, S.; Muhaffel, F.; Riool, M.; Cempura, G.; de Boer, L.; Zaat, S.A.J.; Filemonowicz, A.C.; Cimenoglu, H. Fabrication of Oxide Layer on Zirconium by Micro-Arc Oxidation: Structural and Antimicrobial Characteristics. *Mater. Sci. Eng. C* **2017**, *71*, 565–569. [CrossRef]
37. Durdu, S.; Aktug, S.L.; Aktas, S.; Yalcin, E.; Cavusoglu, K.; Altinkok, A.; Usta, M. Characterization and in Vitro Properties of Antibacterial Ag-Based Bioceramic Coatings Formed on Zirconium by Micro Arc Oxidation and Thermal Evaporation. *Surf. Coat. Technol.* **2017**, *331*, 107–115. [CrossRef]
38. Aktuğ, S.L.; Durdu, S.; Yalçın, E.; Çavuşoğlu, K.; Usta, M. In Vitro Properties of Bioceramic Coatings Produced on Zirconium by Plasma Electrolytic Oxidation. *Surf. Coat. Technol.* **2017**, *324*, 129–139. [CrossRef]
39. Aktug, S.L.; Kutbay, I.; Usta, M. Characterization and Formation of Bioactive Hydroxyapatite Coating on Commercially Pure Zirconium by Micro Arc Oxidation. *J. Alloy. Compd.* **2017**, *695*, 998–1004. [CrossRef]
40. Cengiz, S.; Azakli, Y.; Tarakci, M.; Stanciu, L.; Gencer, Y. Microarc Oxidation Discharge Types and Bio Properties of the Coating Synthesized on Zirconium. *Mater. Sci. Eng. C* **2017**, *77*, 374–383. [CrossRef]
41. Sandhyarani, M.; Rameshbabu, N.; Venkateswarlu, K. Fabrication, Characterization and in-Vitro Evaluation of Nanostructured Zirconia/Hydroxyapatite Composite Film on Zirconium. *Surf. Coat. Technol.* **2014**, *238*, 58–67. [CrossRef]
42. Durdu, S.; Aktug, S.L.; Aktas, S.; Yalcin, E.; Usta, M. Fabrication and in Vitro Properties of Zinc-Based Superhydrophilic Bioceramic Coatings on Zirconium. *Surf. Coat. Technol.* **2018**, *344*, 467–478. [CrossRef]
43. Zhang, L.; Han, Y.; Tan, G. Hydroxyaptite Nanorods Patterned ZrO2 Bilayer Coating on Zirconium for the Application of Percutaneous Implants. *Colloids Surf. B Biointerfaces* **2015**, *127*, 8–14. [CrossRef]
44. Aktug, S.L.; Durdu, S.; Aktas, S.; Yalcin, E.; Usta, M. Characterization and Investigation of in Vitro Properties of Antibacterial Copper Deposited on Bioactive ZrO_2 Coatings on Zirconium. *Thin Solid Film.* **2019**, *681*, 69–77. [CrossRef]
45. Cengiz, S.; Uzunoglu, A.; Huang, S.M.; Stanciu, L.; Tarakci, M.; Gencer, Y. An In-Vitro Study: The Effect of Surface Properties on Bioactivity of the Oxide Layer Fabricated on Zr Substrate by PEO. *Surf. Interfaces* **2021**, *22*, 100884. [CrossRef]

46. Motta, A.T.; Yilmazbayhan, A.; da Silva, M.J.G.; Comstock, R.J.; Was, G.S.; Busby, J.T.; Gartner, E.; Peng, Q.; Jeong, Y.H.; Park, J.Y. Zirconium Alloys for Supercritical Water Reactor Applications: Challenges and Possibilities. *J. Nucl. Mater.* **2007**, *371*, 61–75. [CrossRef]
47. Yau, T.-L.; Annamalai, V.E. Corrosion of Zirconium and its Alloys. In *Reference Module in Materials Science and Materials Engineering*; Elsevier: Amsterdam, The Netherlands, 2016; ISBN 978-0-12-803581-8.
48. Yau, T.-L.; Sutherlin, R.C.; Chang, A.T.I.W. Corrosion of Zirconium and Zirconium Alloys. In *Corrosion: Materials*; Stephen, D.C.; Bernard, S.C., Eds.; ASM International: Ohio, OH, USA, 2018; Chapter 20, pp. 300–324.
49. Chen, Y.; Nie, X.; Northwood, D.O. Investigation of Plasma Electrolytic Oxidation (PEO) Coatings on a Zr–2.5Nb Alloy Using High Temperature/Pressure Autoclave and Tribological Tests. *Surf. Coat. Technol.* **2010**, *205*, 1774–1782. [CrossRef]
50. Cheng, Y.; Wu, F. Plasma Electrolytic Oxidation of Zircaloy-4 Alloy with DC Regime and Properties of Coatings. *Trans. Nonferrous Met. Soc. China* **2012**, *22*, 1638–1646. [CrossRef]
51. Cheng, Y.; Wu, F.; Matykina, E.; Skeldon, P.; Thompson, G.E. The Influences of Microdischarge Types and Silicate on the Morphologies and Phase Compositions of Plasma Electrolytic Oxidation Coatings on Zircaloy-2. *Corros. Sci.* **2012**, *59*, 307–315. [CrossRef]
52. Sreekanth, D.; Rameshbabu, N.; Venkateswarlu, K. Effect of Various Additives on Morphology and Corrosion Behavior of Ceramic Coatings Developed on AZ31 Magnesium Alloy by Plasma Electrolytic Oxidation. *Ceram. Int.* **2012**, *38*, 4607–4615. [CrossRef]
53. Cheng, Y.; Wu, F.; Dong, J.; Wu, X.; Xue, Z.; Matykina, E.; Skeldon, P.; Thompson, G.E. Comparison of Plasma Electrolytic Oxidation of Zirconium Alloy in Silicate- and Aluminate-Based Electrolytes and Wear Properties of the Resulting Coatings. *Electrochim. Acta* **2012**, *85*, 25–32. [CrossRef]
54. Cheng, Y.; Cao, J.; Peng, Z.; Wang, Q.; Matykina, E.; Skeldon, P.; Thompson, G.E. Wear-Resistant Coatings Formed on Zircaloy-2 by Plasma Electrolytic Oxidation in Sodium Aluminate Electrolytes. *Electrochim. Acta* **2014**, *116*, 453–466. [CrossRef]
55. Trivinho-Strixino, F.; da Silva, D.X.; Paiva-Santos, C.O.; Pereira, E.C. Tetragonal to Monoclinic Phase Transition Observed during Zr Anodisation. *J. Solid State Electrochem.* **2013**, *17*, 191–199. [CrossRef]
56. Zou, Z.; Xue, X.; Jia, X.; Du, J.; Wang, R.; Weng, L. Effect of Voltage on Properties of Microarc Oxidation Films Prepared in Phosphate Electrolyte on Zr–1Nb Alloy. *Surf. Coat. Technol.* **2013**, *222*, 62–67. [CrossRef]
57. Xue, W.; Zhu, Q.; Jin, Q.; Hua, M. Characterization of Ceramic Coatings Fabricated on Zirconium Alloy by Plasma Electrolytic Oxidation in Silicate Electrolyte. *Mater. Chem. Phys.* **2010**, *120*, 656–660. [CrossRef]
58. Sukumaran, A.; Sampatirao, H.; Balasubramanian, R.; Parfenov, E.; Mukaeva, V.; Nagumothu, R. Formation of ZrO$_2$–SiC Composite Coating on Zirconium by Plasma Electrolytic Oxidation in Different Electrolyte Systems Comprising of SiC Nanoparticles. *Trans. Inst. Met.* **2018**, *71*, 1699–1713. [CrossRef]
59. Wang, Y.M.; Feng, W.; Xing, Y.R.; Ge, Y.L.; Guo, L.X.; Ouyang, J.H.; Jia, D.C.; Zhou, Y. Degradation and Structure Evolution in Corrosive LiOH Solution of Microarc Oxidation Coated Zircaloy-4 Alloy in Silicate and Phosphate Electrolytes. *Appl. Surf. Sci.* **2018**, *431*, 2–12. [CrossRef]
60. Wu, J.; Lu, P.; Dong, L.; Zhao, M.; Li, D.; Xue, W. Combination of Plasma Electrolytic Oxidation and Pulsed Laser Deposition for Preparation of Corrosion-Resisting Composite Film on Zirconium Alloys. *Mater. Lett.* **2020**, *262*, 127080. [CrossRef]
61. Yilmazbayhan, A.; Motta, A.T.; Comstock, R.J.; Sabol, G.P.; Lai, B.; Cai, Z. Structure of Zirconium Alloy Oxides Formed in Pure Water Studied with Synchrotron Radiation and Optical Microscopy: Relation to Corrosion Rate. *J. Nucl. Mater.* **2004**, *324*, 6–22. [CrossRef]
62. Raj, B.; Mudali, U.K. Materials Development and Corrosion Problems in Nuclear Fuel Reprocessing Plants. *Prog. Nucl. Energy* **2006**, *48*, 283–313. [CrossRef]
63. Fisher, N.J.; Weckwerth, M.K.; Grandison, D.A.E.; Cotnam, B.M. Fretting-Wear of Zirconium Alloys. *Nucl. Eng. Des.* **2002**, *213*, 79–90. [CrossRef]
64. Helmi Attia, M. On the Fretting Wear Mechanism of Zr-Alloys. *Tribol. Int.* **2006**, *39*, 1320–1326. [CrossRef]
65. Cheng, Y.; Xue, Z.; Wang, Q.; Wu, X.-Q.; Matykina, E.; Skeldon, P.; Thompson, G.E. New Findings on Properties of Plasma Electrolytic Oxidation Coatings from Study of an Al–Cu–Li Alloy. *Electrochim. Acta* **2013**, *107*, 358–378. [CrossRef]
66. Martini, C.; Ceschini, L.; Tarterini, F.; Paillard, J.M.; Curran, J.A. PEO Layers Obtained from Mixed Aluminate–Phosphate Baths on Ti–6Al–4V: Dry Sliding Behaviour and Influence of a PTFE Topcoat. *Wear* **2010**, *269*, 747–756. [CrossRef]
67. Yan, Y.; Han, Y.; Huang, J. Formation of Al$_2$O$_3$–ZrO$_2$ Composite Coating on Zirconium by Micro-Arc Oxidation. *Scr. Mater.* **2008**, *59*, 203–206. [CrossRef]
68. Zhang, L.; Zhang, W.; Han, Y.; Tang, W. A Nanoplate-like α-Al$_2$O$_3$ out-Layered Al$_2$O$_3$-ZrO$_2$ Coating Fabricated by Micro-Arc Oxidation for Hip Joint Prosthesis. *Appl. Surf. Sci.* **2016**, *361*, 141–149. [CrossRef]
69. Wei, K.; Chen, L.; Qu, Y.; Yu, J.; Jin, X.; Du, J.; Xue, W.; Zhang, J. Tribological Properties of Microarc Oxidation Coatings on Zirlo Alloy. *Surf. Eng.* **2019**, *35*, 692–700. [CrossRef]
70. Atuchin, V.V.; Aliev, V.S.; Kruchinin, V.N.; Ramana, C.V. Optical Properties of ZrO$_2$ Films Fabricated by Ion Beam Sputtering Deposition at Low Temperature. In Proceedings of the 2007 International Forum on Strategic Technology, Ulaanbaatar, Mongolia, 3–6 October 2007; pp. 529–531.

71. Ramana, C.V.; Utsunomiya, S.; Ewing, R.C.; Becker, U.; Atuchin, V.V.; Aliev, V.S.; Kruchinin, V.N. Spectroscopic Ellipsometry Characterization of the Optical Properties and Thermal Stability of ZrO_2 Films Made by Ion-Beam Assisted Deposition. *Appl. Phys. Lett.* **2008**, *92*, 11917. [CrossRef]
72. Ramana, C.V.; Vemuri, R.S.; Fernandez, I.; Campbell, A.L. Size-Effects on the Optical Properties of Zirconium Oxide Thin Films. *Appl. Phys. Lett.* **2009**, *95*, 231905. [CrossRef]
73. Noor-A-Alam, M.; Ramana, C.; Choudhuri, A. Analysis of microstructure and thermal stability of hafnia-zirconia based thermal barrier coatings. In Proceedings of the 9th Annual International Energy Conversion Engineering Conference, International Energy Conversion Engineering Conference (IECEC), San Diego, CA, USA, 31 July–3 August 2011.
74. Lokesha, H.S.; Nagabhushana, K.R.; Chithambo, M.L.; Singh, F. Down and Upconversion Photoluminescence of ZrO_2:Er^{3+} Phosphor Irradiated with 120 MeV Gold Ions. *Mater. Res. Express* **2020**, *7*, 64006. [CrossRef]
75. Ćirić, A.; Stojadinović, S. Photoluminescence Studies of ZrO2:Tm^{3+}/Yb^{3+} Coatings Formed by Plasma Electrolytic Oxidation. *J. Lumin.* **2019**, *214*, 116568. [CrossRef]
76. Ćirić, A.; Stojadinović, S. Photoluminescence Properties of Pr^{3+} Doped ZrO_2 Formed by Plasma Electrolytic Oxidation. *J. Alloy. Compd.* **2019**, *803*, 126–134. [CrossRef]
77. Stojadinović, S.; Tadić, N.; Vasilić, R. Photoluminescence Properties of Er^{3+}/Yb^{3+} Doped ZrO_2 Coatings Formed by Plasma Electrolytic Oxidation. *J. Lumin.* **2019**, *208*, 296–301. [CrossRef]
78. Stojadinović, S.; Tadić, N.; Vasilić, R. Down- and up-Conversion Photoluminescence of ZrO2:Ho^{3+} and ZrO2:Ho^{3+}/Yb^{3+} Coatings Formed by Plasma Electrolytic Oxidation. *J. Alloy. Compd.* **2019**, *785*, 1222–1232. [CrossRef]
79. Stojadinović, S.; Tadić, N.; Vasilić, R. Down-Conversion Photoluminescence of ZrO2:Er^{3+} Coatings Formed by Plasma Electrolytic Oxidation. *Mater. Lett.* **2018**, *219*, 251–255. [CrossRef]
80. Stojadinović, S.; Tadić, N.; Vasilić, R. Structural and Photoluminescent Properties of ZrO2:Tb^{3+} Coatings Formed by Plasma Electrolytic Oxidation. *J. Lumin.* **2018**, *197*, 83–89. [CrossRef]
81. Stojadinović, S.; Tadić, N.; Vasilić, R. Photoluminescence of Sm^{3+} Doped ZrO_2 Coatings Formed by Plasma Electrolytic Oxidation of Zirconium. *Mater. Lett.* **2016**, *164*, 329–332. [CrossRef]
82. Stojadinović, S.; Vasilić, R.; Radić, N.; Grbić, B. Zirconia Films Formed by Plasma Electrolytic Oxidation: Photoluminescent and Photocatalytic Properties. *Opt. Mater.* **2015**, *40*, 20–25. [CrossRef]
83. Cao, H.; Qiu, X.; Luo, B.; Liang, Y.; Zhang, Y.; Tan, R.; Zhao, M.; Zhu, Q. Synthesis and Room-Temperature Ultraviolet Photoluminescence Properties of Zirconia Nanowires. *Adv. Funct. Mater.* **2004**, *14*, 243–246. [CrossRef]
84. Cong, Y.; Li, B.; Yue, S.; Fan, D.; Wang, X.J. Effect of Oxygen Vacancy on Phase Transition and Photoluminescence Properties of Nanocrystalline Zirconia Synthesized by the One Pot Reaction. *J. Phys. Chem. C* **2009**, *113*, 13974–13978. [CrossRef]
85. Kumari, L.; Li, W.Z.; Xu, J.M.; Leblanc, R.M.; Wang, D.Z.; Li, Y.; Guo, H.; Zhang, J. Controlled Hydrothermal Synthesis of Zirconium Oxide Nanostructures and Their Optical Properties. *Cryst. Growth Des.* **2009**, *9*, 3874–3880. [CrossRef]
86. Jain, N.; Singh, R.K.; Sinha, S.; Singh, R.A.; Singh, J. Color Tunable Emission through Energy Transfer from Yb^{3+} Co-Doped SrSnO3:Ho^{3+} Perovskite Nano-Phosphor. *Appl. Nanosci.* **2018**, *8*, 1267–1278. [CrossRef]

Article

Ce^{3+}/Eu^{2+} Doped Al_2O_3 Coatings Formed by Plasma Electrolytic Oxidation of Aluminum: Photoluminescence Enhancement by $Ce^{3+} \rightarrow Eu^{2+}$ Energy Transfer

Stevan Stojadinović * and Aleksandar Ćirić

Faculty of Physics, University of Belgrade, Studentski trg 12-16, 11000 Belgrade, Serbia; aleksandar.ciric@ff.bg.ac.rs
* Correspondence: sstevan@ff.bg.ac.rs; Tel.: +381-11-7158161

Received: 16 November 2019; Accepted: 25 November 2019; Published: 3 December 2019

Abstract: Plasma electrolytic oxidation (PEO) of aluminum in electrolytes containing CeO_2 and Eu_2O_3 powders in various concentrations was used for creating Al_2O_3 coatings doped with Ce^{3+} and Eu^{2+} ions. Phase and chemical composition, surface morphology, photoluminescence (PL) properties and energy transfer from Ce^{3+} to Eu^{2+} were investigated. When excited by middle ultraviolet radiation, $Al_2O_3:Ce^{3+}/Eu^{2+}$ coatings exhibited intense and broad emission PL bands in the ultraviolet/visible spectral range, attributed to the characteristic electric dipole $4f^05d^1 \rightarrow 4f^1$ transition of Ce^{3+} (centered at about 345 nm) and $4f^65d^1 \rightarrow 4f^7$ transition of Eu^{2+} (centered at about 405 and 500 nm). Due to the overlap between the PL emission of $Al_2O_3:Ce^{3+}$ and the PL excitation of $Al_2O_3:Eu^{2+}$, energy transfer from Ce^{3+} sensitizer to the Eu^{2+} activator occurs. The energy transfer is identified as an electric dipole–dipole interaction. The critical distance between Eu^{2+} and Ce^{3+} ions in Al_2O_3 was estimated to be 8.6 Å by the spectral overlap method.

Keywords: plasma electrolytic oxidation; Al_2O_3; energy transfer; photoluminescence; Ce^{3+}/Eu^{2+}

1. Introduction

Ce^{3+} and Eu^{2+} ions have found widespread use as activators in luminescent materials because their parity allows $4f^1 \rightarrow 4f^05d^1$ and $4f^7 \rightarrow 4f^65d^1$ electric dipole transitions, respectively. Outer orbitals at 5d states of Ce^{3+} and Eu^{2+} ions are sensitive to crystal-field splitting and nephelauxetic effects [1,2], therefore, energy positions of excitation and emission bands of these ions can be tuned by changing the host matrix. Ce^{3+} ion also acts as an effective sensitizer due to its effective absorption [3]. Numerous studies demonstrated the importance and methods for enhancing the photoluminescence (PL) via an efficient energy transfer (e.g., [4,5]). The energy transfer from the Ce^{3+} sensitizer to the Eu^{2+} activator has been reported for some Ce^{3+}/Eu^{2+} doped host lattices because of the spectral overlap between Ce^{3+} emission band and Eu^{2+} excitation band [6–9]. The aim of this work is to investigate PL properties of Ce^{3+}/Eu^{2+} doped Al_2O_3 coatings created by plasma electrolytic oxidation (PEO) of pure aluminum and an energy transfer from Ce^{3+} to Eu^{2+} ions.

Recent investigations have shown that PEO process of aluminum is capable of preparation of Ce^{3+} and Eu^{2+} doped Al_2O_3 coatings in electrolyte enriched with CeO_2 and Eu_2O_3 particles, respectively [10,11]. As ionic radii of Al^{3+} (0.53 Å), and Eu^{3+} (0.95 Å) and Ce^{4+} (0.97 Å) [12] significantly differ, the substitution of Al^{3+} by either Eu^{3+} or Ce^{4+} produces oxygen vacancies that in turn distort the crystal lattice. The excess O^{2-} ions give their electrons to the lattice ($2O^{2-} - 4e^- \rightarrow O_2$), which are then captured by Ce^{4+} and Eu^{3+}, leading to their reduction to Ce^{3+} and Eu^{2+}, respectively [13].

Thus, the interactions of CeO_2 or Eu_2O_3 particles and Al_2O_3 coatings via the PEO process cause the reduction of Ce^{4+} to Ce^{3+} ions and Eu^{3+} to Eu^{2+} ions, enabling their efficient incorporation into Al_2O_3. Our assumption was that Al_2O_3:Ce^{3+}/Eu^{2+} coatings could be formed if the PEO of aluminum is performed with an electrolyte containing CeO_2 and Eu_2O_3 particles.

2. Materials and Methods

Pure aluminum foil (99.9%) with 0.25 mm thickness was used as a substrate in PEO. Before PEO, the substrate was cut into plates (25 × 10 mm^2), cleaned ultrasonically in acetone and ethanol, rinsed with distilled water, dried in a warm air stream and sealed with insulation resin leaving a working area of ~15 × 10 mm^2. PEO was carried out in a double-walled glass cell, with water cooling. The temperature of the electrolyte was kept at the constant (20 ± 1) °C during the PEO. Aluminum samples were used as an anode while the cathode was a cylinder of stainless steel. An aqueous solution of 0.1 M H_3BO_3 + 0.05 M $Na_2B_4O_7$ was employed as a supporting electrolyte. CeO_2 and Eu_2O_3 particles in different concentrations were added to the electrolyte. The magnetic stirrer allowed for the homogeneous particle distribution in the glass cell. The coatings were formed using a constant current density regime of 150 mA/cm^2 for 10 min. Following PEO, the samples were rinsed in distilled water and dried in warm air.

The surface morphology of formed coatings was examined by scanning electron microscope (SEM, Tokyo, Japan) JEOL 840A and surface elemental distribution were determined by the integrated energy-dispersive x-ray spectroscopy (EDS, Oxford, UK). Phase composition of created PEO coatings was analyzed by X-ray diffraction (XRD, Tokyo, Japan), using a Rigaku Ultima IV diffractometer over a scanning range of 20°–80° with a 0.05° step size and 2°/min scanning speed. The Kratos AXIS Supra photoelectron spectrometer using a monochromatic Al K$_\alpha$ source with the energy of 1486.6 eV was used for obtaining the X-ray photoelectron spectra (XPS). The base pressure in the analysis chamber was 5×10^{-8} Pa.

Room temperature PL spectral measurements were acquired by a Horiba Jobin Yvon Fluorolog FL3-22 spectrofluorometer (Edison, NJ, USA), with a 450 W xenon lamp as the excitation source, coupled to a double grating monochromator in a wavelength range 220–600 nm. A 290–850 nm double grating monochromator and a side-on Hamamatsu 928P photomultiplier tube (PMT) allowed for recording of the PL emission spectra. The spectra were corrected for the spectral distribution of the excitation lamp and the measuring system's spectral response.

3. Results and Discussion

3.1. PEO Coatings' Morphology, Phase and Chemical Composition

PEO is a high-voltage anodizing process of some metals (Al, Mg, Ta, Ti, etc.) and their alloys, combined with the plasma formation, as indicated by the presence of micro-discharges on the metal surface, and followed by the gas evolution [14]. High pressure (~10^2 MPa) and temperature (10^3 K to 10^4 K) at the micro-discharge sites induce numerous processes (electrochemical, thermodynamical, and plasma-chemical) responsible for structural, compositional, and morphological modifications of the obtained oxide coatings. The PEO formed oxide coatings usually have crystalline phases with constituents originating from both the electrolyte and the metal. During the PEO of aluminum, Al_2O_3 layer grows at the oxide/electrolyte and aluminum/oxide interfaces as a consequence of a strong electric field (~10^7 V/cm) induced relocation of O^{2-}/OH^- and Al^{3+} ions across the oxide [15]. Moreover, melted aluminum and electrolyte components get drawn into the micro-discharge channels. From the channels, the oxidized aluminum gets ejected to the coating surface contacting the electrolyte. In that manner, the coating thickness increases around the channels. During the PEO, the electrophoretic effect drives CeO_2 and Eu_2O_3 particles towards the anode. Local temperature at the micro-discharge sites is higher than the melting points of CeO_2 and Eu_2O_3 particles (~2400 °C), enabling the molten particles to react with Al_2O_3 and form Ce and Eu ions doped Al_2O_3 coatings.

SEM images of coating surfaces, formed in supporting electrolyte by PEO, with the addition of CeO_2 and Eu_2O_3 particles in various concentrations, are shown in Figure 1a. PEO coating displays a porous morphology with pores of varying diameter and shape, which appear at the sites of the micro-discharge channels because of the gas evolution erupting the molten oxide material during the PEO process [16]. The surface morphology of the formed coatings does not significantly change with the addition of CeO_2 and Eu_2O_3 particles in supporting electrolyte. Elemental mapping by EDS shows that the main constituents of the coatings are uniformly distributed O, Ce, and Eu from the electrolyte and Al from the substrate (Figure 1b).

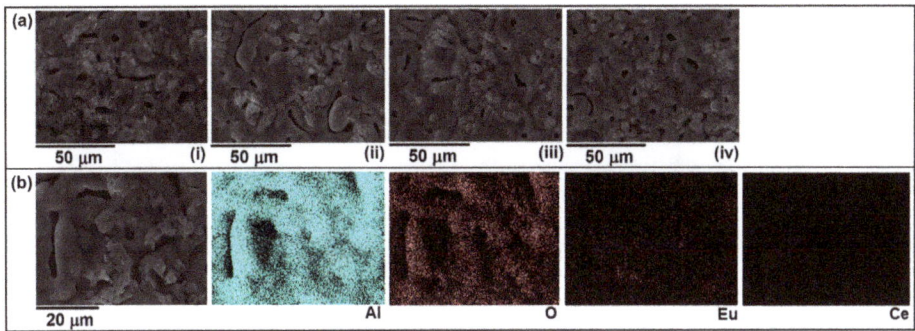

Figure 1. (a) SEM images of coatings created in supporting electrolyte: (i) Without addition CeO_2 and Eu_2O_3 and with the addition of (ii) 4 g/L CeO_2, (iii) 4 g/L Eu_2O_3, (iv) 4 g/L CeO_2 + 4 g/L Eu_2O_3. (b) Energy-dispersive x-ray spectroscopy (EDS) maps of coating created in supporting electrolyte with the addition of 4 g/L CeO_2 + 4 g/L Eu_2O_3.

Integrated EDS analysis of all coatings created in the experiment is given in Table 1. The content of Ce and Eu in the coatings increases with increased concentration of CeO_2 and Eu_2O_3 particles in supporting electrolyte, respectively.

Table 1. EDS analysis of plasma electrolytic oxidation (PEO) coatings coalesced in supporting electrolyte with different concentrations of added CeO_2 and Eu_2O_3 powders.

Concentration of Particles in Supporting Electrolyte	Atomic (%)			
	O	Al	Ce	Eu
4 g/L CeO_2	65.25	34.62	0.13	-
4 g/L Eu_2O_2	65.19	34.55	-	0.26
4 g/L CeO_2 + 0.5 g/L Eu_2O_3	64.89	34.95	0.10	0.06
4 g/L CeO_2 + 1 g/L Eu_2O_3	64.73	35.06	0.11	0.09
4 g/L CeO_2 + 2 g/L Eu_2O_3	64.39	35.35	0.11	0.16
4 g/L CeO_2 + 4 g/L Eu_2O_3	64.79	34.82	0.12	0.27
0 g/L CeO_2 + 0.5 g/L Eu_2O_3	64.95	34.98	-	0.07
4 g/L CeO_2 + 0.5 g/L Eu_2O_3	64.90	34.94	0.10	0.06
8 g/L CeO_2 + 0.5 g/L Eu_2O_3	64.99	34.73	0.21	0.07
12 g/L CeO_2 + 0.5 g/L Eu_2O_3	64.97	34.65	0.32	0.06

XRD patterns of created coatings are presented in Figure 2. Diffraction peaks corresponding to the gamma phase of Al_2O_3 (reference ICCD card No. 10-0425) and Al substrate (reference ICCD card No. 89-4037) are observed in XRD patterns. Obviously, the formation of gamma Al_2O_3 during the PEO is favored by the rapid solidification of molten alumina due to the contact with surrounding low-temperature electrolyte [17]. We were not able to detect any peaks corresponding to CeO_2 or Eu_2O_3 particles as well as any other Ce or Eu species, probably due to low concentration of incorporated Ce and Eu elements into Al_2O_3 coatings (Table 1).

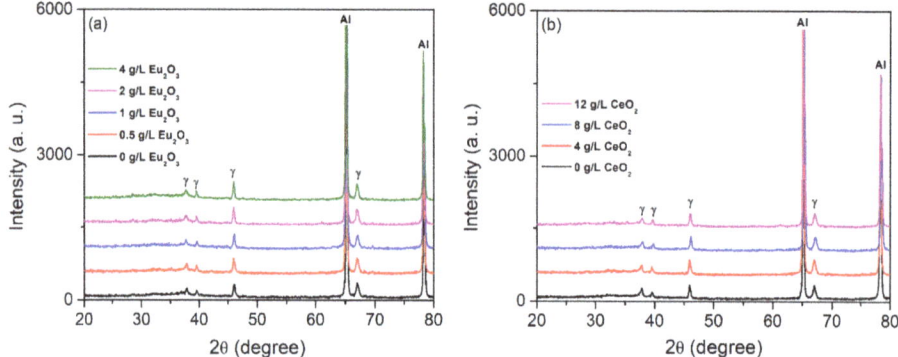

Figure 2. (**a**) XRD patterns of coatings created in 0.1 M H_3BO_3 + 0.05 M $Na_2B_4O_7$ + 4 g/L CeO_2 with the addition Eu_2O_3 in different concentrations, (**b**) XRD patterns of coatings formed in 0.1 M H_3BO_3 + 0.05 M $Na_2B_4O_7$ + 0.5 g/L Eu_2O_3 with the addition CeO_2 in different concentrations.

For further investigation of the chemical nature and oxidation state of Ce and Eu in created coatings, we recorded a high-resolution Ce 3d and Eu 3d core-level spectra of Ce and Eu doped Al_2O_3 coating created in supporting electrolyte with the addition of 4 g/L CeO_2 + 4 g/L Eu_2O_3 particles (Figure 3). The peak deconvolution of the high-resolution XPS peaks of Ce 3d shows that the level of Ce ion is composed of four Gaussian peaks, i.e., two pairs of doublets ($3d_{5/2}$ at 881.5 eV and 885.3 eV, $3d_{3/2}$ at 899.5 eV and 903.9 eV), characteristic for Ce^{3+} oxidation state [18]. The Eu 3d core level spectrum consists of two doublets: (i) The Eu $3d_{5/2}$ (at 1135.3 eV) and Eu $3d_{3/2}$ (at 1165.3 eV) peaks, attributed to Eu^{3+} oxidation state, and (ii) at lower binding energies at ca. 1125.1 eV and 1155.2 eV, from the Eu^{2+} oxidation state [19]. These results indicate that interaction between CeO_2 and Eu_2O_3 particles with Al_2O_3, under environmental conditions set by PEO, causes the reduction of Ce^{4+} ions to Ce^{3+} ions and the reduction of some of Eu^{3+} to Eu^{2+} ions in Al_2O_3 host (e.g., [10,11] and references therein).

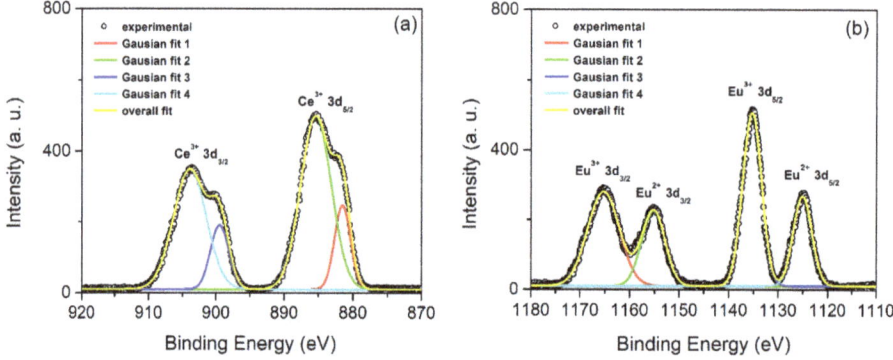

Figure 3. (**a**) High-resolution XPS spectrum of Ce 3d, (**b**) high-resolution Eu 3d XPS spectrum of coating created with the addition of 4 g/L CeO_2 + 4 g/L Eu_2O_3 to the supporting electrolyte.

3.2. PL of Al_2O_3:Ce^{3+} and Al_2O_3:Eu^{2+} Coatings

The PL excitation and emission spectra of Ce^{3+} and Eu^{2+} singly doped Al_2O_3 coatings are presented in Figure 4. The PL excitation spectrum of Al_2O_3:Ce^{3+} coating (Figure 4a) exhibited the broad band in the range from 250 to 340 nm and peaked at about 285 nm, corresponding to the direct excitation from the Ce^{3+} ground state 4f to the field-splitting levels of 5d state [20]. The corresponding PL emission spectrum excited at 285 nm shows one broad emission band peaked at about 340 nm, attributed to the transitions of 5d excited state to the $^2F_{7/2}$ and $^2F_{5/2}$ ground states [20].

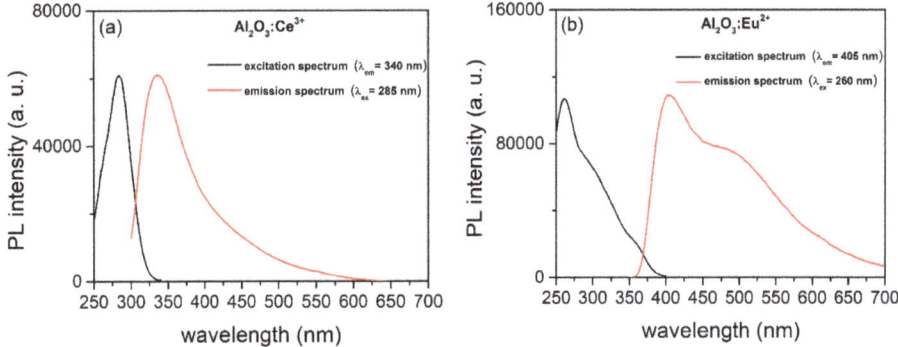

Figure 4. Photoluminescence (PL) excitation and emission spectra of: (**a**) Al_2O_3:Ce^{3+}, (**b**) Al_2O_3:Eu^{2+}, created in supporting electrolyte with the addition of 4 g/L CeO_2 and 4 g/L Eu_2O_3, respectively.

Figure 4b presents PL excitation and emission spectra of Al_2O_3 coating doped with Eu ions. The PL spectra show the characteristic Eu^{2+} broad band excitation and emission [21]. Although XPS indicates that Eu incorporated into Al_2O_3 is also in the 3+ oxidation state, typical f-f transitions of Eu^{3+} ions have not been identified in PL excitation and emission spectra, not even under the 395 nm excitation corresponding to the $^7F_0 \rightarrow {}^5L_6$ transition of Eu^{3+}. The PL excitation spectrum shows a large absorption band ranging from 250 to 330 nm, with center at 260 nm, attributed to the $4f^65d^1$ multiplet excited states of Eu^{2+} ions. Under the excitation of 260 nm the PL emission spectrum shows two distinct bands reaching maxima at about 405 and 500 nm, caused by the $4f^65d^1 \rightarrow 4f^7$ ($^8S_{7/2}$) transition of Eu^{2+} ions. This transition is structurally sensitive to the local environment around the Eu^{2+} in Al_2O_3 [21]. The gamma Al_2O_3 has two different sites for Al^{3+} ions. The appearance of two emission bands for a single transition of Eu^{2+} in Al_2O_3:Eu^{2+} is thus attributed to the Eu^{2+} substituting the Al^{3+} ions at both crystallographic sites.

3.3. PL of Al_2O_3:Ce^{3+}/Eu^{2+} Coatings

Figure 4 shows the spectral overlapping of the Ce^{3+} PL emission to the Eu^{2+} PL excitation in Al_2O_3 between 300 and 400 nm, which indicates that the energy transfer from a sensitizer Ce^{3+} to an activator Eu^{2+} is possible. To verify the energy transfer from Ce^{3+} to Eu^{2+} in Al_2O_3:Ce^{3+}/Eu^{2+} coatings, the PL emission spectra excited at 260 and 285 nm of coatings formed in supporting electrolyte with the addition of 4 g/L CeO_2 and different concentrations of Eu_2O_3 are shown in Figure 5. The PL emission spectra of Al_2O_3:Ce^{3+}/Eu^{2+} coatings consist of the emission band peaking at about 345 nm, assigned to the $4f^05d^1 \rightarrow 4f^1$ transition of Ce^{3+} ion, and two emission bands peaking at about 405 and 500 nm, attributed to the $4f^65d \rightarrow 4f^7$ transition of Eu^{2+} ion. The intensity of the PL band of Ce^{3+} decreases with the increasing concentration of Eu_2O_3 in supporting electrolyte, i.e., content of incorporated Eu in Al_2O_3 coatings (Table 1), but the intensity of PL bands of Eu^{2+} increases. With the higher Eu^{2+} doping content, the Ce^{3+} emission practically disappears, and only the Eu^{2+} emission remains in the PL spectra of Al_2O_3:Ce^{3+}/Eu^{2+} coatings. These results indicate that the energy transfer from Ce^{3+} to Eu^{2+} is confirmed. The intensity of PL bands of Eu^{2+} not only increases due to $Ce^{3+} \rightarrow Eu^{2+}$ energy transfer, but also due to the increase of the Eu content in the Al_2O_3 coatings.

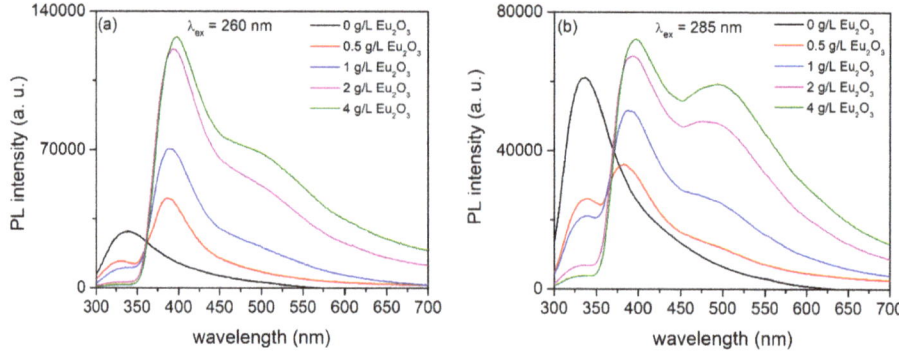

Figure 5. PL emission spectra of $Al_2O_3:Ce^{3+}/Eu^{2+}$ coatings formed in supporting electrolyte with the addition of 4 g/L CeO_2 and different concentrations of Eu_2O_3 excited at: (**a**) 260 nm, (**b**) 285 nm.

$Ce^{3+} \rightarrow Eu^{2+}$ energy transfer in $Al_2O_3:Ce^{3+}/Eu^{2+}$ coatings created in supporting electrolyte with the addition of 0.5 g/L Eu_2O_3 and different concentrations of CeO_2 is verified as well (Figure 6). The intensity of PL bands originating from Ce^{3+} and Eu^{2+} increases with increasing concentration of CeO_2 in the supporting electrolyte, i.e., content of incorporated Ce^{3+} in Al_2O_3 coatings (Table 1). The intensity of PL bands of Eu^{2+} increases due to $Ce^{3+} \rightarrow Eu^{2+}$ energy transfer, but the intensity of the PL band of Ce^{3+} increases due to the increase of the Ce^{3+} content in the Al_2O_3 coatings.

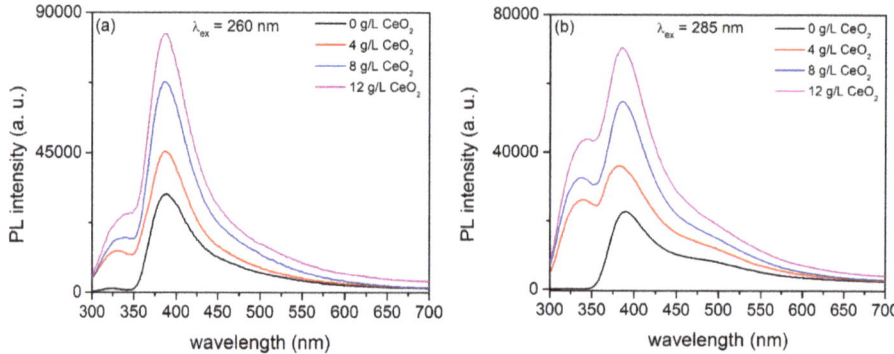

Figure 6. PL emission spectra of $Al_2O_3:Ce^{3+}/Eu^{2+}$ coatings formed in supporting electrolyte with the addition of 0.5 g/L Eu_2O_3 and different concentrations of CeO_2 excited at: (**a**) 260 nm, (**b**) 285 nm.

The energy transfer efficiency (η_T) from Ce^{3+} to Eu^{2+} can be calculated by the formula $\eta_T = 1 - I_S/I_{S0}$, where I_S and I_{S0} are the PL intensities of the Ce^{3+} emissions with and without the presence of Eu^{2+}, respectively [22]. The energy transfer efficiencies from Ce^{3+} to Eu^{2+} are calculated from the spectra in Figure 5 and presented in Figure 7. The energy transfer efficiency increases with increasing Eu^{2+} concentration, indicating that the energy transfer from Ce^{3+} to Eu^{2+} is effective under middle UV excitation. The energy transfer $Ce^{3+} \rightarrow Eu^{2+}$ can also be observed via the shortening of the Ce^{3+} emission decay times with the increasing Eu^{2+} concentration, as demonstrated in Ref. [7].

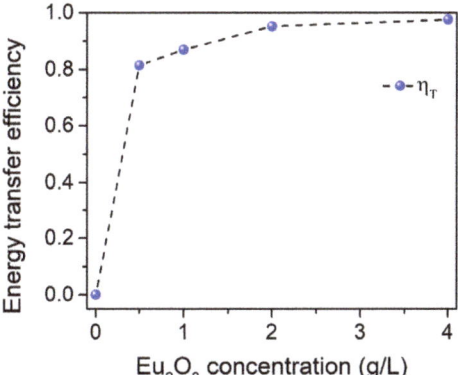

Figure 7. Energy transfer efficiency of $Al_2O_3:Ce^{3+}/Eu^{2+}$ coatings created in supporting electrolyte with the addition of 4 g/L CeO_2 and different concentrations of Eu_2O_3.

In order to identify the energy transfer mechanism from a Ce^{3+} sensitizer to an Eu^{2+} activator, the equation of exchange interaction and electric multipolar interactions proposed by Dexter and Reisfeld was used [3]:

$$\frac{I_{S0}}{I_S} \propto C^{\frac{n}{3}}. \qquad (1)$$

From Equation (1) the dominant multipolar interaction can be identified, where C is the total concentration of Ce^{3+} and Eu^{2+} ions, and $n = 6, 8, 10$ values correspond to dipole–dipole, dipole–quadropole and quadropole–quadropole interactions, respectively. In Figure 8, the plots of $C^{n/3}$ vs. I_{S0}/I_S are presented. The best fit is with $n = 6$, i.e., the energy transfer $Ce^{3+} \rightarrow Eu^{2+}$ is primarily due to the electric dipole–dipole interaction.

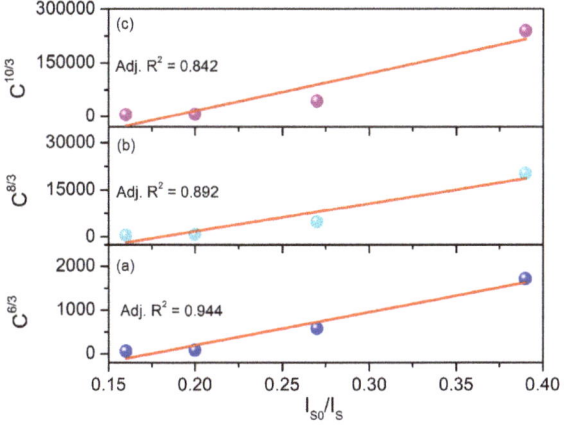

Figure 8. Plot of $C^{n/3}$ vs. I_{S0}/I_S for: (**a**) $n = 6$, (**b**) $n = 8$, (**c**) $n = 10$.

The critical transfer distance for electric dipole–dipole interactions equal to [8]:

$$R_C^6 [\text{Å}] = 6 \cdot \frac{10^3}{L^4} \cdot \frac{\int I(Ce) \cdot I(Eu) dE}{\int I(Ce) dE \cdot \int I(Eu) dE}, \qquad (2)$$

where $I(Ce)$ and $I(Eu)$ are the intensities of Ce^{3+} emission and Eu^{2+} excitation spectra, respectively, and L is the largest energy point at which the spectral overlap occurs. As can be extracted from the graph in Figure 9, that point is equal to 3.985 eV, the term with integrals is equal to 0.3596 eV^{-1}, and thus the critical distance was estimated to be 8.6 Å.

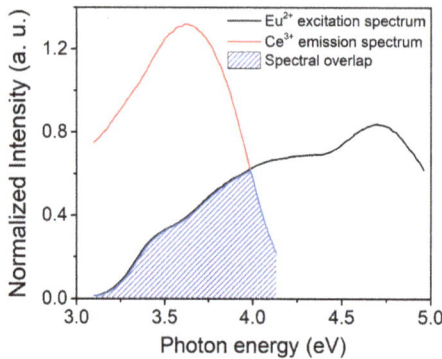

Figure 9. Normalized spectral line-shapes for the Ce^{3+} emission and Eu^{2+} excitation.

3.4. CIE Chromaticity Analysis

The Commission International de l'Eclairage (CIE) 1931 (x,y) coordinates of Al_2O_3:Ce^{3+}/Eu^{2+} coatings created in supporting electrolyte with the addition different concentration of CeO_2 and Eu_2O_3 were estimated by the JOES software, v. 2.8 [23], and the results are presented in Figure 10 and Table 2. From the diagram, the evident is the shift from pure blue towards the white color by increasing Eu^{2+} concentration. There is also a small shift in the same direction by changing the excitation from 260 to 285 nm. The emission color of the samples with constant Eu^{2+} is almost independent on the Ce^{3+} concentration. Thus, the emission color of Al_2O_3:Ce^{3+}/Eu^{2+} can be fine-tuned from pure blue to white by increasing Eu^{2+} concentration.

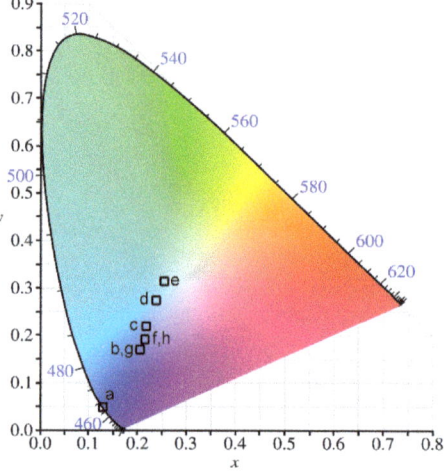

Figure 10. Commission International de l'Eclairage (CIE) 1931 chromaticity diagram of Al_2O_3:Ce^{3+}/Eu^{2+} coatings doped with various concentrations of CeO_2 and Eu_2O_3 particles and excited at 260 nm. Labels are explained in Table 2.

Table 2. CIE x, y coordinates, color purity and dominant wavelength of Al_2O_3:Ce^{3+}/Eu^{2+}, coatings doped with various concentrations of CeO_2 and Eu_2O_3 particles and excited at 260 and 285 nm.

Label	Concentration of Particles		λ_{ex} (nm)	x	y	Color Purity	$\lambda_{dominant}$ (nm)
	CeO_2	Eu_2O_3					
a	4	0	260	0.10644	0.08603	1	467
b	4	0.5	260	0.20944	0.18067	0.606957	471
c	4	1	260	0.21295	0.21447	0.501821	477
d	4	2	260	0.24027	0.25217	0.374879	483
e	4	4	260	0.25693	0.27666	0.285429	488
f	0	0.5	260	0.21503	0.19208	0.550572	472
g	8	0.5	260	0.2022	0.17773	0.606957	471
h	12	0.5	260	0.21743	0.19386	0.550572	472

4. Conclusions

In summary, we have successfully prepared Al_2O_3 coatings doped with Ce^{3+} and Eu^{2+} ions by the PEO process and their PL properties have been investigated in detail. PL emission spectra of Al_2O_3:Ce^{3+}/Eu^{2+} are a sum of PL originating from 5d–4f transitions of Ce^{3+} and Eu^{2+} ions. Due to the spectral overlapping of the Ce^{3+} PL emission with the Eu^{2+} PL excitation in Al_2O_3, an energy transfer from Ce^{3+} sensitizer to Eu^{2+} activator occurs. The energy transfer mechanism is dominant by electric dipole–dipole interaction.

Author Contributions: S.S. conceptualized the idea; S.S. developed the coating; S.S. developed the methodology; S.S. and A.Ć. performed the experiments, and analyzed data; S.S. and A.Ć. contributed equally in manuscript preparation, editing and submission.

Funding: This research was funded by the Ministry of Education, Science and Technological Development of the Republic of Serbia under project No. 171035 and by the European Union Horizon 2020 research and innovation program under the Marie Sklodowska-Curie grant agreement No. 823942.

Conflicts of Interest: The authors declare no conflict of interest.

References

1. Li, G.G.; Tian, Y.; Zhao, Y.; Lin, J. Recent progress in luminescence tuning of Ce^{3+} and Eu^{2+}-activated phosphors for pc- WLEDs. *Chem. Soc. Rev.* **2015**, *44*, 8688–8713. [CrossRef] [PubMed]
2. Yan, S. On the origin of temperature dependence of the emission maxima of Eu^{2+} and Ce^{3+}- activated phosphors. *Opt. Mater.* **2018**, *79*, 172–185. [CrossRef]
3. Li, K.; Shang, M.; Lian, H.; Lin, J. Recent development in phosphors with different emitting colors via energy transfer. *J. Mater. Chem. C* **2016**, *4*, 5507–5530. [CrossRef]
4. Martín-Cano, D.; Martín-Moreno, L.; García-Vidal, F.J.; Moreno, E. Resonance Energy Transfer and Superradiance Mediated by Plasmonic Nanowaveguides. *Nano Lett.* **2010**, *10*, 3129–3134. [CrossRef]
5. Caligiuri, V.; Palei, M.; Imran, M.; Manna, L.; Krahne, R. Planar Double-Epsilon-Near-Zero Cavities for Spontaneous Emission and Purcell Effect Enhancement. *ACS Photonics* **2018**, *5*, 2287–2294. [CrossRef]
6. Zhao, J.; Sun, X.; Wang, Z. Ce^{3+}/Eu^{2+} doped $SrSc_2O_4$ phosphors: Synthesis, luminescence and energy transfer from Ce^{3+} to Eu^{2+}. *Chem. Phys. Lett.* **2018**, *691*, 68–72. [CrossRef]
7. Hu, Z.; Cheng, Z.; Dong, P.; Zhang, H.; Zhang, Y. Enhanced photoluminescence property of single-component $CaAlSiN_3$:Ce^{3+}, Eu^{2+} multicolor phosphor through Ce^{3+}-Eu^{2+} energy transfer. *J. Alloys Compd.* **2017**, *727*, 633–641. [CrossRef]
8. Yan, J.; Liu, C.; Zhou, W.; Huang, Y.; Tao, Y.; Liang, H. VUV-UV–vis photoluminescence of Ce^{3+} and Ce^{3+}-Eu^{2+} energy transfer in $Ba_2MgSi_2O_7$. *J. Lumin.* **2017**, *185*, 251–257. [CrossRef]
9. Li, C.; Zheng, H.; Wei, H.; Qiu, S.; Xu, L.; Wang, X.; Jiao, H. A color tunable and white light emitting $Ca_2Si_5N_8$:Ce^{3+}, Eu^{2+} phosphor via efficient energy transfer for near-UV white LEDs. *Dalton Trans.* **2018**, *47*, 6860–6867. [CrossRef]
10. Stojadinović, S.; Vasilić, R. Eu^{2+} photoluminescence in Al_2O_3 coatings obtained by plasma electrolytic oxidation. *J. Lumin.* **2018**, *199*, 240–244. [CrossRef]

11. Stojadinović, S.; Vasilić, R. Photoluminescence of Ce^{3+} and Ce^{3+}/Tb^{3+} ions in Al_2O_3 host formed by plasma electrolytic oxidation. *J. Lumin.* **2018**, *203*, 576–581. [CrossRef]
12. Kim, D.-J. Lattice Parameters, Ionic Conductivities, and Solubility Limits in Fluorite-Structure MO_2 Oxide [M = Hf^{4+}, Zr^{4+}, Ce^{4+}, Th^{4+}, U^{4+}] Solid Solutions. *J. Am. Ceram. Soc.* **1989**, *72*, 1415–1421. [CrossRef]
13. Xie, H.; Lu, J.; Guan, Y.; Huang, Y.; Wei, D.; Seo, H.J. Abnormal Reduction, $Eu^{3+} \rightarrow Eu^{2+}$, and Defect Centers in Eu^{3+}-Doped Pollucite, $CsAlSi_2O_6$, Prepared in an Oxidizing Atmosphere. *Inorg. Chem.* **2014**, *53*, 827–834. [CrossRef] [PubMed]
14. Stojadinović, S.; Vasilić, R.; Perić, M. Investigation of plasma electrolytic oxidation on valve metals by means of molecular spectroscopy–a review. *RSC Adv.* **2014**, *4*, 25759–25789. [CrossRef]
15. Snizhko, L.O.; Yerokhin, A.L.; Pilkington, A.; Gurevina, N.L.; Misnyankin, D.O.; Leyland, A.; Matthews, A. Anodic processes in plasma electrolytic oxidation of aluminium in alkaline solutions. *Electrochim. Acta* **2004**, *49*, 2085–2095. [CrossRef]
16. Yerokhin, A.L.; Nie, X.; Leyland, A.; Matthews, A.; Dowey, S.J. Plasma electrolysis for surface engineering. *Surf. Coat. Technol.* **1999**, *122*, 73–93. [CrossRef]
17. McPherson, R. Formation of metastable phases in flame and plasma-prepared alumina. *J. Mater. Sci.* **1973**, *8*, 851–858. [CrossRef]
18. Yagoub, M.Y.A.; Swart, H.C.; Noto, L.L.; Bergman, P.; Coetsee, E. Surface characterization and photoluminescence properties of Ce^{3+},Eu co-doped SrF_2 nanophosphor. *Materials* **2015**, *8*, 2361–2375. [CrossRef]
19. Rogers, J.J.; MacKenzie, K.J.D.; Rees, G.; Hanna, J.V. New phosphors based on the reduction of Eu(III) to Eu(II) in ion-exchanged aluminosilicate and gallium silicate inorganic polymers. *Ceram. Int.* **2018**, *44*, 1110–1119. [CrossRef]
20. Wang, F.; Wang, W.; Zhang, L.; Zheng, J.; Jin, Y.; Zhang, J. Luminescence properties and its red shift of blue emitting phosphor $Na_3YSi_3O_9:Ce^{3+}$ for UV LED. *RSC Adv.* **2017**, *7*, 27422–27430. [CrossRef]
21. Yang, Y.; Wei, H.; Zhang, L.; Kisslinger, K.; Melcher, C.L.; Wu, Y. Blue emission of Eu^{2+}-doped translucent alumina. *J. Lumin.* **2015**, *168*, 297–303. [CrossRef]
22. Chen, H.; Li, C.; Hua, Y.; Yu, L.; Jiang, Q.; Deng, D.; Zhao, S.; Ma, H.; Xu, S. Influence of energy transfer from Ce^{3+} to Eu^{2+} on luminescence properties of $Ba_3Si_6O_9N_4:Ce^{3+},Eu^{2+}$ phosphors. *Ceram. Int.* **2014**, *40*, 1979–1983. [CrossRef]
23. Ćirić, A.; Stojadinović, S.; Sekulić, M.; Dramićanin, M.D. JOES: An application software for Judd-Ofelt analysis from Eu^{3+} emission spectra. *J. Lumin.* **2019**, *205*, 351–356. [CrossRef]

© 2019 by the authors. Licensee MDPI, Basel, Switzerland. This article is an open access article distributed under the terms and conditions of the Creative Commons Attribution (CC BY) license (http://creativecommons.org/licenses/by/4.0/).

Article

Infrared Absorption Study of Zn–S Hybrid and ZnS Ultrathin Films Deposited on Porous AAO Ceramic Support

Maksymilian Włodarski [1], Matti Putkonen [2] and Małgorzata Norek [3],*

[1] Institute of Optoelectronics, Military University of Technology, Str. gen. Sylwestra Kaliskiego 2, 00-908 Warsaw, Poland; maksymilian.wlodarski@wat.edu.pl
[2] Department of Chemistry, University of Helsinki, P.O. Box 55, FI-00014 Helsinki, Finland; matti.putkonen@helsinki.fi
[3] Institute of Materials Science and Engineering, Military University of Technology, Str. gen. Sylwestra Kaliskiego 2, 00-908 Warsaw, Poland
* Correspondence: malgorzata.norek@wat.edu.pl

Received: 21 April 2020; Accepted: 6 May 2020; Published: 9 May 2020

Abstract: Infrared (IR) spectroscopy is a powerful technique to characterize the chemical structure and dynamics of various types of samples. However, the signal-to-noise-ratio drops rapidly when the sample thickness gets much smaller than penetration depth, which is proportional to wavelength. This poses serious problems in analysis of thin films. In this work, an approach is demonstrated to overcome these problems. It is shown that a standard IR spectroscopy can be successfully employed to study the structure and composition of films as thin as 20 nm, when the layers were grown on porous substrates with a well-developed surface area. In contrast to IR spectra of the films deposited on flat Si substrates, the IR spectra of the same films but deposited on porous ceramic support show distinct bands that enabled reliable chemical analysis. The analysis of Zn-S ultrathin films synthesized by atomic layer deposition (ALD) from diethylzinc (DEZ) and 1,5-pentanedithiol (PDT) as precursors of Zn and S, respectively, served as proof of concept. However, the approach presented in this study can be applied to analysis of any ultrathin film deposited on target substrate and simultaneously on porous support, where the latter sample would be a reference sample dedicated for IR analysis of this film.

Keywords: ultrathin films; infrared spectroscopy; detection limit; ZnS; atomic layer deposition (ALD); molecular layer deposition (MLD)

---

1. Introduction

Thin-film materials are becoming increasingly important in many technological fields, such as electronics, optics and biotechnology [1–3]. The broad application of thin films has become possible thanks to the constant development of deposition techniques which have enabled the fabrication of thin films with controllable thickness, composition and structure. One of the most advanced methods to fabricate thin films is atomic layer deposition (ALD), which is distinguished from other techniques by its high conformability. Thanks to the self-limiting reaction at the surface between gaseous precursor molecules and chemical groups at a substrate [4,5], uniform layers can be grown on a high aspect ratio and three dimensionally structured materials. The ALD technique allows the film thickness and architecture to be controlled down to the molecular level. Nowadays, ultrathin films play a major role in devices such as microelectronic components, solar cells, LED displays, and sensors [6–9]. ALD-like surface-limiting reactions are also applied for organic compounds, enabling the molecular layer deposition (MLD) of polymers and hybrid inorganic thin films [10]. In ultrathin films (thickness ranging from ca. 1 to 100 nm) the physical and chemical properties of the surface and interfaces are strongly enhanced compared to the bulk materials. In these systems, the phenomena such as quantum size effect and/or tunnelling effect for electron transfer [11] and shorter diffusion length for carriers [12], changes in transition temperature [13], crystallization [14], anisotropy of thermal conductivity [15] can occur. These properties are, in turn, strictly related to the crystal structure and chemical composition of ultrathin-film materials. Therefore, characterizing the structural parameters of thin films is of high significance, both scientifically and technologically.

Infrared (IR) absorption spectroscopy is a powerful technique, commonly applied to characterize the chemical structure and dynamics of almost all types of sample (liquids, powders, films). It is based on the absorption of IR light by the vibration states of molecules when the energy difference between the vibrations states matches that of the incident light. The absorption peaks, with a specific position usually within the 400–4000 cm^{-1} range, are called the "chemical fingerprints" of the molecules. Moreover, the technique is non-invasive, label-free (no special sample preparation is needed), and is able to quickly obtain desired structural information, e.g., types of functional groups within a sample, with high spectroscopic precision. However, the IR spectroscopy is more effective at longer wavelengths since the penetration depth (d_p) is proportional to the wavelength. This principle can generate serious problems in the measurement of very thin films, because the signal-to-noise-ratio drops rapidly when the sample thickness gets much smaller than the d_p. To address those problems, advanced methods and instrumentations were developed, including infrared reflection–absorption spectroscopy (IRRAS) [16,17], polarization modulation-IRRAS (PM-IRRAS) [18,19], the application of special filters such as a non-scattering metal grid [20] or, quite recently, photothermal nanomechanical IR sensing (NAM-IR) [21,22], which push infrared spectroscopic techniques to new limits. These techniques are, however, expensive, time consuming, and usually require complicated sample preparation or post-processing treatments. Hence, there is still a need for a simple, reliable, and nondestructive method to analyze thin films by standard IR spectroscopy without the necessity of using the sophisticated technical approaches.

Here, we present a simple approach to study ultrathin (20–60 nm) films by standard IR spectroscopy. The approach relies on using a ceramic support with a nanoporous structure with a high specific surface area to increase the signal-to-noise-ratio. The ceramic support was fabricated by aluminum anodization. As an example, the chemical and structural properties of Zn–S ultrathin films, prepared by low-temperature ALD from diethylzinc and 1,5-pentanedithiol (PTD), are studied [23]. It was shown previously that the low deposition temperature results in amorphous ZnS films with high carbon content [24]. It is, therefore, expected that difunctional 1,5-pentanedithiol can react at the surface, leading to MLD-like films [10]. However, there are no published results of the IR spectroscopic data of these films. In this work, long and parallel channels of anodic alumina (AAO) were uniformly covered by the Zn–S thin films, thus enhancing the effective optical path length and the interaction of the Zn–S film with incident light. As a result, the Zn–S layers as thin as 20 nm were successfully analyzed by a standard IR spectroscopy. The IR spectral response from the same layers but deposited on a flat support was below the detection limit. Moreover, it was observed that the signal-to-noise ratio increases as the thickness of the Zn–S layer increases, and as the diameter of the AAO channels and interpore distance increase. The approach can be applied to study any thin film deposited on a target surface and, at the same time, on porous support, where the latter sample would be used as a reference sample to study the chemical behavior and structure of this thin film by standard IR spectroscopy, without the need to use sophisticated IR spectrometers or technical approaches, which are not always readily available.

2. Materials and Methods

AAO porous substrates were fabricated by aluminum (Al) anodization. High-purity Al foil (99.9995% Al, Goodfellow), with a thickness of about 0.25 mm, was cut into rectangular specimens (2 cm × 1 cm). Before the anodization process, the Al foils were degreased in acetone and ethanol and subsequently electropolished in a 1:4 mixture of 60% HClO$_4$ and ethanol at 0 °C, constant voltage of 25 V, for 2.5 min. Next, the samples were rinsed with distilled water, ethanol and dried. As prepared Al specimens were insulated at the back and the edges with acid-resistant tape, and serve as the anode. A Pt grid was used as a cathode and the distance between both electrodes was kept constant (ca. 5 cm). A large, 1 L electrochemical cell and cooling bath thermostat (model MPC-K6, Huber company, Offenburg, Germany) were employed in the anodizing process. An adjustable DC power supply with a voltage range of 0–300 V and current range of 0–5 A, purchased from NDN, model GEN750_1500 TDK Lambda, TDK Co. Tokyo, Japan, was used to control the applied voltage. The AAO_phosphoric substrates were prepared using a hard anodization (HA) method at voltage 150 V in a 0.3 M H$_2$C$_2$O$_4$ ethanol-modulated solution with 4:1 v/v water to EtOH, at 0 °C. The samples were pre-anodized at 40 V for 5–8 min prior to the application of a given voltage. Then, the voltage was slowly increased to a target value (150 V) at a rate ranging from around 0.04 to 0.06 V/s, and the samples were anodized for 2 h. Alumina was chemically removed using a mixture of 6 wt.% phosphoric acid and 1.8 wt.% chromic acid at 60 °C for 120 min. The second anodization, which was performed in 0.1 M phosphoric acid solution with 1:4 v/v mixture of ethanol and water as a solvent, for 1 h and at 0 °C under the same anodizing voltage as used in the first step. The AAO_oxalic substrates were prepared by two-step anodization in 0.3 M H$_2$C$_2$O$_4$ water-based solution at temperature of 35 °C for 2 h (both first and second anodization). To obtain the geometrical parameters of the fabricated samples, Fast Fourier transforms (FFTs) were generated based on three SEM images taken at the same magnification for every sample, and were further used in calculations with WSxM software (version 5.0) [25]. The average interpore distance (D_c) was estimated as an inverse of the FFT's radial average abscissa from three SEM images for each sample. The average pore diameter (D_p) was estimated from three SEM images for each analyzed sample, using NIS-Elements software provided by Nikon Company, Tokyo, Japan.

The ZnS thin films were deposited by ALD using Picosun SUNALE™ R-200 ALD reactor (Picosun Oy, Espoo, Finland) in a single wafer mode. Depositions were carried out at 150 °C with diethylzinc (DEZ, Strem Chemicals >95%) and 1,5-pentanedithiol (PDT, Merck 96%) as precursors. DEZ was kept at 20 °C, whereas PDT was evaporated from Picohot 200 hot source and held at 55 °C in order to obtain sufficient vapor pressure. N$_2$ (purity 99.999%) from liquid nitrogen gas (LNG) was used as a carrier gas. The ALD deposition cycle consisted of DEZ pulse/purge/PDT pulse/purge with 0.2/4/0.3/4 s timing, respectively. Zn–S film deposition rate was 0.09 Å/cycle when measured after 500–2500 deposition cycles, resulting in film thicknesses of ca. 20, 40 and 60 nm. The ZnS films were deposited on AAO porous substrates and on reference silicon (Si) (Siltronic AG, Münich, Germany). Thicknesses of the deposited films were measured by ex-situ ellipsometry (Sentech SE400adv, SENTECH Instruments GmbH, Berlin, Gemany).

Microanalysis of chemical composition was made using a field-emission scanning electron microscope FE-SEM (AMETEK, Inc., Montvale, NJ, USA) equipped with energy dispersive X-ray spectrometer (EDS). The chemical composition analysis was performed at 20 kV, magnification of 500, spot of 2.0, and with a constant distance of samples to the detector (WD = 10). Each measurement was repeated three times and an average of the three measurements was taken to determine the chemical composition of a studied samples.

Coated AAO was subjected to an XRD phase and structural analysis using a Rigaku Ultima IV diffractometer (Co-Kα λ = 179,003 Å) (Rigaku, Tokyo, Japan) with operating parameters of 40 mA and 40 kV in a continuous mode, with a speed of 1 deg/min. Parallel beam geometry was used together with a fast linear detector (DeteX Ultra). The slit sizes were kept constant during the investigation (fixed slit mode). The phase identification of the base structure was performed with PDXL (Rigaku, version 2.8.4.0) software and the PDF4 database.

Fourier-transform infrared spectroscopy (FTIR) spectra were recorded using Spectrum GX Optica spectrometer from Perkin-Elmer, Waltham, MA, USA with diffuse reflectance accessory. Light scattered from surface of the sample was collected in a full π steradian angle. The angle of incidence was

38°. For films deposited on Si substrate, clean Si was used as a 100% reflectance reference. For films deposited on AAO transmissive substrate, aluminum sheet was used as a 100% reflectance reference.

3. Results and Discussion

Two groups of samples were analyzed by a standard IR spectroscopy: the Zn–S layers directly after the ALD process (before annealing), and the same samples after annealing. The samples were annealed under argon (99.999%) atmosphere at 400 °C for 1 h. The annealing conditions were chosen based on our previous work [24] to crystallize the amorphous ZnS material. The ZnS layers with the thicknesses of 20, 40, and 60 nm (ZnS–20 nm, ZnS–40 nm, and ZnS–60 nm, respectively) were deposited on flat Si and on porous AAO substrates with an interpore distance (D_c) of ~346 nm (AAO-phosphoric). The 20 and 40 nm thick ZnS layers were also deposited on the porous AAO substrates characterized by about three times smaller D_c ~ 120 nm (AAO-oxalic). The 60-nm thick ZnS thin film was not deposited on the AAO_oxalic because, in this case, the pore diameter was too small (D_p ~ 90 nm) to allow for a uniform coverage of the AAO channels by the Zn–S material. In Figure 1, SEM images of the top view of the samples before annealing are demonstrated. The deposition of ZnS films causes the gradual filling of pores in both AAO_phosphoric and AAO_oxalic porous substrates. As a result, the pore diameter observed in the SEM images decreases with increasing thickness of the ZnS films.

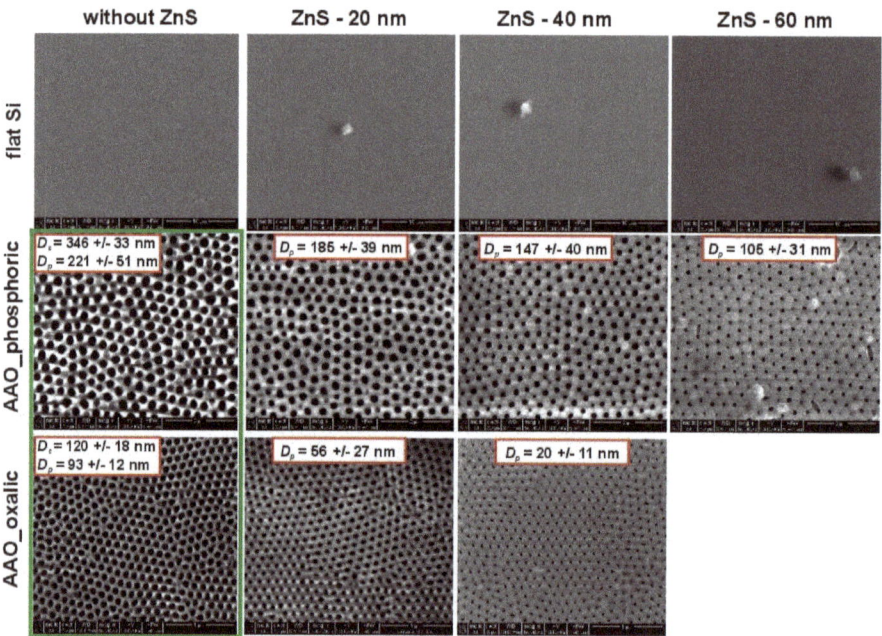

Figure 1. SEM images of the ZnS layers with a thickness of 20, 40, and 60 nm deposited on flat Si substrates (first row; scale bar = 10 μm), porous anodic alumina (AAO) substrates with D_c ~ 346 nm (second row; scale bar = 2 μm), and the porous AAO substrates with D_c ~ 120 nm (third row, scale bar = 1 μm).

In Figure 2a, an SEM image of a cross-sectional view of the AAO-phosphoric substrate covered with the 60 nm thick Zn–S layer before annealing (AAO_phosphoric-ZnS-60 nm sample), together with EDS elemental mapping images of Al, O, Zn, S, and C (K lines), are presented. The images demonstrate that Zn, S, and C elements are evenly distributed along the entire pore cross-section. After the dissolution of AAO, bundles of ZnS nanotubes were obtained (Figure 2b), which further confirms the homogeneous and complete coverage of the interior of the AAO channels by ZnS material during ALD process. In the EDS spectrum shown in Figure 2c, there are peaks from Zn, S, and C (O and Si elements come from NOA 61 optical adhesive by which the ZnS nanotubes were attached to glass [26]), but no peak from Al was recorded, proving a selective and total dissolution of anodic alumina (in 0.1 M NaOH solution) in the AAO_phosphoric-ZnS-60 nm sample.

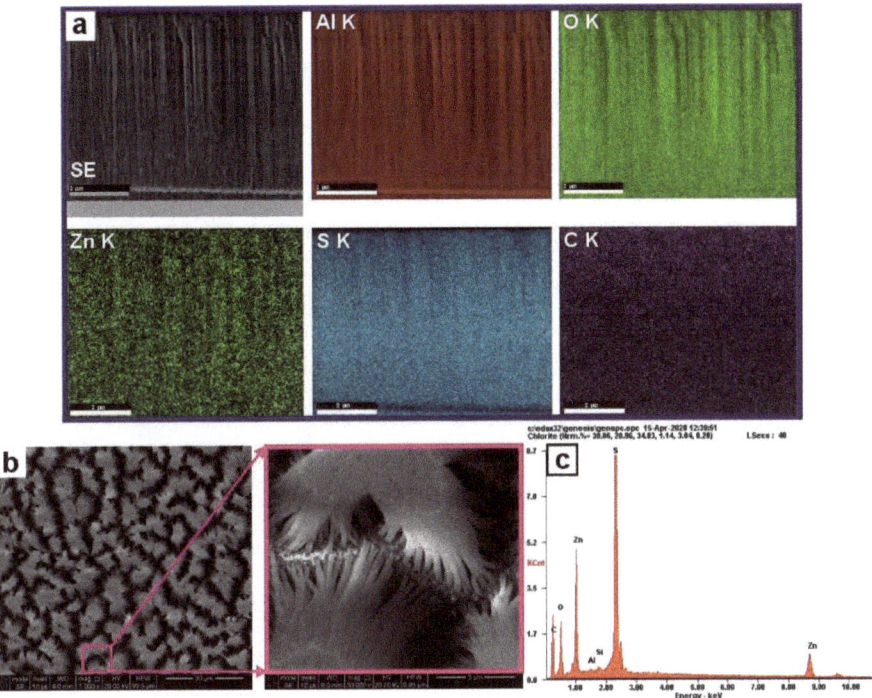

Figure 2. SE image of a cross-sectional view of the AAO_phosphoric-ZnS-60 nm sample before annealing, and corresponding EDS elemental mapping of Al, O, Zn, S, and C (scale bar = 2 µm) (a); SEM image of the ZnS nanotubes after dissolution of AAO in the AAO_phosphoric-ZnS-60 nm sample (b) and corresponding EDS spectrum (c).

In Figure 3, the EDS spectra of the AAO_phosphoric-ZnS-60 nm sample before (a) and after annealing (b) are shown (the spectra were acquired until comparable total number of counts). The spectra demonstrate an apparent increase in Zn lines in the expense of the C line after annealing process. A collective EDS elemental analysis of the AAO_phosphoric-ZnS-20 nm, AAO_phosphoric-ZnS-40 nm, AAO_phosphoric-ZnS-60 nm samples before and after annealing is demonstrated in Figure 3c. The graphs in Figure 3c present the atomic concentration (at.%) of all elements as a function of the ZnS layer thickness. It can be noticed that the concentration of Zn, S, and C elements evidently increases with the increase in the ZnS layer thickness, except the C content in the samples after annealing, which stays on a similar level (~3 at.%). As the amount of all elements adds up to 100% in the EDS analysis, the at.% concentration of all other elements (Al, O, and P; P is present in the AAO matrix due to the incorporation of PO_4^{3-} ions during anodization in 0.1 M H_3PO_4 electrolyte [27,28]) tends to decrease with the increase in Zn, S, and C elements. As observed previously [24], before annealing the amount of S is about two times higher than the amount of Zn. The higher amount of S is accompanied by a relatively large amount of C (from ca. 9 to about 14 at.% for the increasing ZnS film thickness) in the samples before annealing. We observed earlier by TOF-ERDA that as-deposited films had a 1:2 Zn:S ratio and 51% C and 27% of H [24]. After annealing, the amount of Zn and S was equalized to similar values and the amount of C dropped significantly to about 3 at.% for all samples. The large amount of carbon, hydrogen and 1:2 Zn:S suggest the presence of unreacted organic precursor in as-deposited film, indicating MLD-type growth of inorganic-organic hybrid materials, while a significant decrease in C in the samples after annealing indicates the removal of the carbon-containing species. Moreover, almost stoichiometric composition of the Zn and S obtained after annealing may indicate the crystallization of the amorphous ZnS material. In order to find out more information about the composition and structure of the ZnS thin films before and after the annealing process, the material was further investigated by XRD, and subsequently by IR spectroscopy.

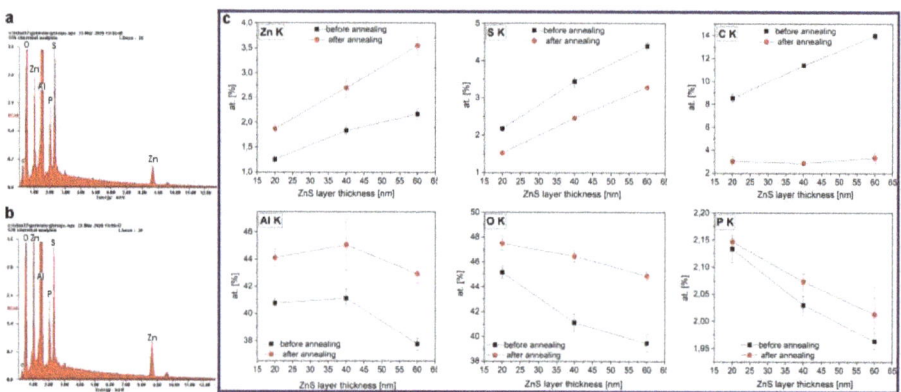

Figure 3. EDS spectrum of the AAO_phosphoric-ZnS-60 nm sample before (**a**) and after annealing (**b**); EDS elemental analysis of the samples AAO_phosphoric-ZnS-20 nm, AAO_phosphoric-ZnS-40 nm, AAO_phosphoric-ZnS-60 nm before and after annealing (**c**).

In Figure 4, the XRD patterns of the AAO_phosphoric-ZnS-20 nm, AAO_phosphoric-ZnS-40 nm, AAO_phosphoric-ZnS-60 nm samples after annealing are demonstrated. As shown in the example of the AAO_phosphoric-ZnS-60 nm sample, the ZnS material was amorphous before the annealing process (directly after ALD). After the process, the ZnS crystallizes into cubic form with clearly visible reflections from (111), (220), and (311) crystallographic planes (PDF Number 01-072-4841). The thicker the ZnS film, the more intensive are the peaks in the XRD patterns. A similar transformation of the ZnS material upon heating was observed in the ZnS layer deposited on a flat Si substrate [24]. However, as compared to the XRD profiles of the latter films, the peaks in the Figure 4 are broader. For instance, the FWHM of the (111) reflection is ~6.1° in the XRD pattern of the 60 nm-thick ZnS layer deposited on an AAO_phosphoric porous substrate, while the full width at half maximum (FWHM) of the (111) reflection is around 2.2° in the XRD pattern of the corresponding ZnS layer deposited on the non-porous Si substrate. The peak broadening can be due to smaller crystal size and greater microstrain induced by a porous substrate. In the XRD pattern of the AAO_phosphoric-ZnS-20 nm sample, the lines corresponding to Al are also visible (the AAO was not detached from Al foil after anodization), most probably due to a slightly different sample positioning in the diffractometer and/or larger penetration depth of the X-ray beam, owing to the smaller ZnS film thickness, and thus lower material density.

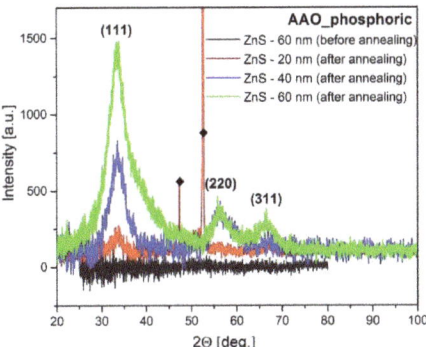

Figure 4. XRD patterns of the AAO_phosphoric-ZnS-60 nm sample before annealing and the samples AAO_phosphoric-ZnS-20 nm, AAO_phosphoric-ZnS-40 nm, AAO_phosphoric-ZnS-60 nm after annealing (the black diamond symbols signify the reflections originating from the Al substrate).

In Figure 5, the IR spectrum of the studied samples are presented. The IR spectra of the amorphous ZnS layers deposited on flat Si substrates display no peaks: the signal was below the detection limit of IR spectroscopy (Figure 5a). In the IR spectra of the same ZnS layers but deposited on porous AAO substrates (Figure 5b,c), distinct absorption peaks are visible, which enable chemical and structural analysis of the layers. A broad vibration centered at ~3450–3650 cm^{-1}, present both in Figure 5b,c, is due to the stretching vibrations of O–H bond (v(OH)) of adsorbed water [29–33]. The vibration confirms the presence of moisture on the surface of ZnS thin films. In the spectral range extending from about 3000 to ca. 1300 cm^{-1}, there are many absorption peaks which cannot originate from chemical groups in the AAO substrate (since they are absent in the IR spectrum of pure AAO in Figure 5b). Zn–S ultrathin films were prepared by low-temperature ALD from diethylzinc (DEZ) and 1,5-pentanedithiol (PDT) as precursors of Zn and S, respectively. Characteristic peaks of IR spectra for 1,5-pentanedithiol can be attributed from each as-deposited Zn–S films on AAO indicating MLD type growth [34,35]. The strongest dips at 2927 and 2851 cm^{-1} were ascribed before to the CH$_2$ asymmetric (v_a(CH$_2$)) and symmetric (v_s(CH$_2$)) stretching vibrational modes, respectively [29,30,36–38]. These dips are also the most intensive ones in the IR spectrum of pure PDT [34,35] and can thus be regarded as chemical fingerprints of this compound. Clearly, with the increase in the thickness of the ZnS layers, the intensity of those two dips increases (Figure 5b). The dips at 1454, 1433 and 1365 cm^{-1} can be ascribed to the CH or CH$_2$ bending modes [29–31,39]. It can be thus concluded that IR analysis provided evidence for the presence of Zn–S-pentanedithiol type MLD hybrid materials, which was already envisaged based on the EDS analysis. Apart from the bands that can be associated with PDT chemical groups, in the IR spectra there are dips that originate from other functional groups. For instance, a weak band at 2361 cm^{-1} can be attributed to the C=O stretching mode (v_3(CO$_2$)) of absorbed CO$_2$ [32,36,40]. The bands at 1559 and 1586 cm^{-1} were previously associated with C=O stretching vibrations of the absorbed CO$_2$ as well [29,41,42], but could also come from the products of the DEZ and PDT surface reaction. Moreover, there are two dips in the 900–1200 cm^{-1} spectral range that are also present in the IR spectrum of the pure AAO_phosphoric sample. Both dips are characteristic for phosphonates, which display bands due to the P–O stretching vibration [43,44]. The dip at ~1002 cm^{-1} can be assigned to the symmetric vibration of the P–O chemical bond (v_sP–O) of PO$_3^{2-}$ group, whereas the one at 1164 cm^{-1} corresponds most probably to the P=O vibrations. These bands confirm the incorporation of phosphonate ions into the AAO framework during anodization (see the EDS analysis in Figure 3c). After ZnS deposition, the dips tend to shift towards a lower energy (enlarged part of the IR spectra in Figure 5b, marked by dotted magenta frame). A weak and broad band appearing at around 484 cm^{-1} may originate from the Zn–O stretching vibration [33,36,37,45].

In the ZnS thin films deposited on the AAO_oxalic (Figure 5c), only the bands at 2927 and 2851 cm^{-1}, characteristic for PTD, are clearly visible, while the other bands (at 1454, 1433 and 1365 cm^{-1}) are missing. Moreover, the bands at 2927 and 2851 cm^{-1} are much weaker in the ZnS film deposited on AAO_oxalic with smaller D_c and D_p parameters as compared to the ones deposited on AAO_phosphoric. Besides, in the IR spectra there are other peaks related to the AAO porous ceramic anodized in oxalic acid solution [46,47]. The bands at 2339 and 2263 cm^{-1} in pure AAO_oxalic sample may come from v(CH) vibrations of the absorbed hydrogen [48]. The peak at around 1044 cm^{-1} can be assigned to the coupling of the C–C stretching vibration and the O–C=O bending vibration [49].

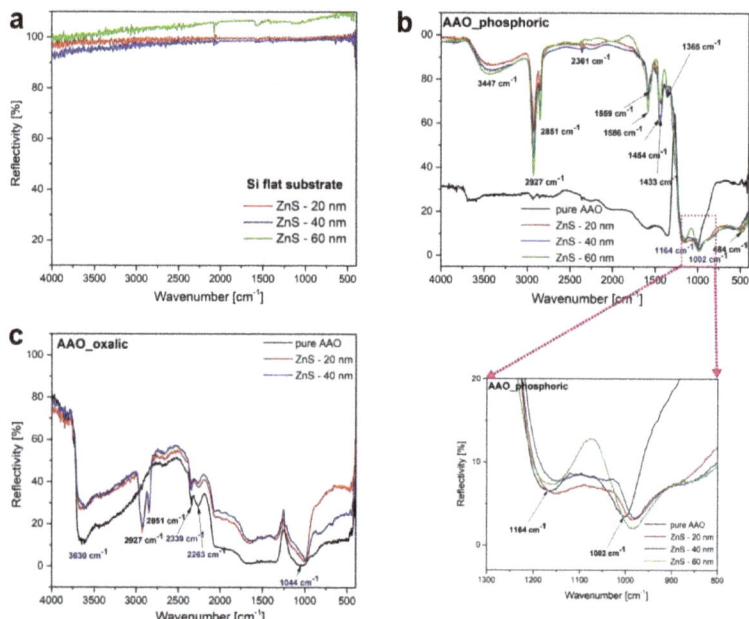

Figure 5. IR spectrum of the Zn–S-1,5-pentanedithiol layers deposited on flat Si substrate (**a**) and on AAO_phosphoric (**b**) and AAO_oxalic (**c**) porous substrates before annealing process.

After annealing, the structure of the ZnS films changes substantially (Figure 6). The peaks that were related to the 1,5-pentanedithiol have gone or weakened significantly (dotted, squared frames in Figure 6a,b). On the other hand, the peaks that were ascribed to chemical groups in AAO porous substrates are preserved in the IR spectra. Likewise, a weak and broad peak at around 484 cm^{-1} that was associated with Zn–O stretching vibration, indicating a partial oxidation of surface Zn. The IR analysis confirmed thus that the low-temperature ALD/MLD process resulted in the synthesis of amorphous Zn–S-1,5-pentanedithiol hybrid materials. Heating of the layers at 400 °C for 1h is sufficient to improve the composition and structure of ZnS. The analysis of the same ZnS layers deposited on flat Si substrate was not possible by a standard IR spectroscopy due to the constraints related to small layer thickness. Long and parallel channels of anodic alumina (AAO), uniformly covered by the ZnS thin films, enhanced the effective optical path length and the interaction of the ZnS film with incident IR light, thus enabling a reliable chemical analysis of the ZnS layers as thin as 20 nm.

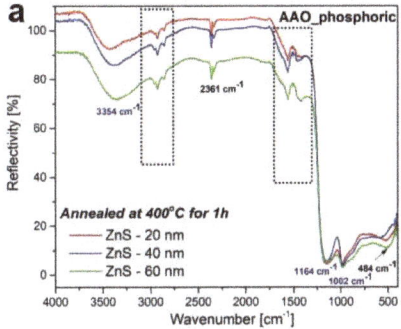 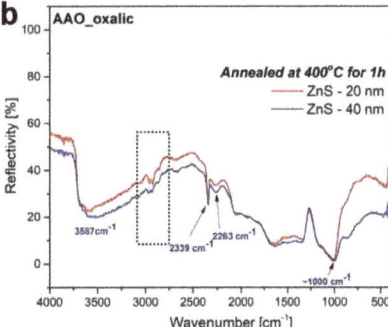

Figure 6. IR spectrum of the ZnS layers deposited on AAO_phosphoric (**a**) and AAO_oxalic (**b**) porous substrates after annealing at 400 °C for 1 h.

4. Conclusions

In this work, it was demonstrated that ultrathin films (20–60 nm) can be successfully investigated by standard IR spectroscopy when deposited on porous substrates with a well-developed surface area. As an example, hybrid Zn–S-1,5-pentanedithiol and ZnS thin films were analyzed which were deposited on flat Si and porous AAO substrates. In contrast to the films deposited on the porous AAO substrates, the same films on a flat Si substrate could not be analyzed by IR spectroscopy: in the IR spectrum, no bands were detected. Thanks to the application of the AAO porous support, it was possible to study the structure and composition of the Zn–S-1,5-pentanedithiol and ZnS ultrathin films synthesized by ALD/MLD from diethylzinc (DEZ) and 1,5-pentanedithiol (PDT) as precursors of Zn and S, respectively. According to IR analysis, it was revealed that the low temperature ALD deposition of 150 °C resulted in the production of amorphous hybrid material with 1,5-pentanethiol. Moreover, it was observed that the signal-to-noise ratio increases as the thickness of the Zn–S layer increases, and as the diameter of the AAO channels and interpore distance increase. After annealing at 400 °C for 1 h, the amorphous films transformed into cubic ZnS form, which was also indicated in the IR spectra: the bands related to the unreacted precursors disappeared or weakened substantially. The approach presented in this study can be applied to study any ultrathin films, where the porous ceramic support can serve as a reference substrate for the film to be studied by a standard IR spectroscopy.

Author Contributions: Conceptualization, M.N.; methodology, M.P.; investigation, M.W. and M.P.; measurements, M.W. and M.N.; writing—original draft preparation, M.N.; writing—review and editing, M.P.; All authors have read and agreed to the published version of the manuscript.

Funding: The work was supported by the statutory research funds of the Department of Functional Materials and Hydrogen Technology, Military University of Technology, Warsaw, Poland. M.P. acknowledges funding from the Academy of Finland by the profiling action on Matter and Materials, Grant No. 318913.

Conflicts of Interest: The authors declare no conflict of interest.

References

1. Wetzig, K.; Schneider, C.M. *Metal Based Thin Films for Electronics*, 2nd ed.; Wiley-VCH Verlag GmbH & Co. KGaA: Weinheim, Germany, 2006.
2. Piegari, A.; Flory, F. *Optical Thin Films and Coatings: From Materials to Applications*, 2nd ed.; Woodhead Publishing: Cambridge, UK, 2018.
3. Grandin, G.M.; Textor, M. *Intelligent Surfaces in Biotechnology: Scientific and Engineering Concepts, Enabling Technologies, and Translation to Bio-Oriented Applications*; John Wiley & Sons, Inc.: Hoboken, NJ, USA, 2012.
4. Johnson, R.W.; Hultqvist, A.; Bent, S.F. A brief review of atomic layer deposition: From fundamentals to applications. *Mater. Today* **2014**, *17*, 236–246. [CrossRef]

5. Knez, M.; Nielsch, K.; Niinisto, L. Synthesis and Surface Engineering of Complex Nanostructures by Atomic Layer Deposition. *Adv. Mater.* **2007**, *19*, 3425–3438. [CrossRef]
6. Wang, K.X.; Yu, Z.; Liu, V.; Cui, Y.; Fan, S. Absorption Enhancement in Ultrathin Crystalline Silicon Solar Cells with Antireflection and Light-Trapping Nanocone Gratings. *Nano Lett.* **2012**, *12*, 1616–1619. [CrossRef] [PubMed]
7. Islam, R.; Saraswat, K. Limitation of Optical Enhancement in Ultra-thin Solar Cells Imposed by Contact Selectivity. *Sci. Rep.* **2018**, *8*, 8863. [CrossRef] [PubMed]
8. Tchoe, Y.; Chung, K.; Lee, K.; Jo, J.; Chung, K.; Hyun, J.K.; Kim, M.; Yi, G.-C. Free-standing and ultrathin inorganic light-emitting diode array. *NPG Asia Mater.* **2019**, *11*, 377. [CrossRef]
9. Youngquist, R.C.; Nurge, M.A.; Fisher, B.H.; Malocha, D.C. A Resistivity Model for Ultrathin Films and Sensors. *IEEE Sens. J.* **2014**, *15*, 2412–2418. [CrossRef]
10. Sundberg, P.; Karppinen, M. Organic and inorganic–organic thin film structures by molecular layer deposition: A review. *Beilstein J. Nanotechnol.* **2014**, *5*, 1104–1136. [CrossRef]
11. Zhao, J.; Bradbury, C.R.; Huclova, S.; Potapova, I.; Carrara, M.; Fermin, D.J. Nanoparticle-Mediated Electron Transfer Across Ultrathin Self-Assembled Films. *J. Phys. Chem. B* **2005**, *109*, 22985–22994. [CrossRef]
12. Tang, C.; Yan, Z.; Wang, Q.; Chen, J.; Zhu, M.; Liu, B.; Liu, F.; Sui, C. Ultrathin amorphous silicon thin-film solar cells by magnetic plasmonic metamaterial absorbers. *RSC Adv.* **2015**, *5*, 81866–81874. [CrossRef]
13. Koh, Y.P.; McKenna, G.B.; Simon, S.L. Calorimetric glass transition temperature and absolute heat capacity of polystyrene ultrathin films. *J. Polym. Sci. Part B Polym. Phys.* **2006**, *44*, 3518–3527. [CrossRef]
14. Harada, K.; Sugimoto, T.; Kato, F.; Watanabe, K.; Matsumoto, Y. Thickness dependent homogeneous crystallization of ultrathin amorphous solid water films. *Phys. Chem. Chem. Phys.* **2020**, *22*, 1963–1973. [CrossRef] [PubMed]
15. Yang, W.; Zhao, Z.; Wu, K.; Huang, R.; Liu, T.; Jiang, H.; Chen, F.; Fu, Q. Ultrathin flexible reduced graphene oxide/cellulose nanofiber composite films with strongly anisotropic thermal conductivity and efficient electromagnetic interference shielding. *J. Mater. Chem. C* **2017**, *5*, 3748–3756. [CrossRef]
16. Hoffmann, F.M.; Levinos, N.J.; Perry, B.N.; Rabinowitz, P. High-resolution infrared reflection absorption spectroscopy with a continuously tunable infrared laser: CO on Ru(001). *Phys. Rev. B* **1986**, *33*, 4309–4311. [CrossRef] [PubMed]
17. Jiang, E.Y. *Advanced FT-IR Spectroscopy*; Thermo Electron Corporation: Madison, WI, USA, 2003; p. 58.
18. Wang, C.; Zheng, J.; Zhao, L.; Rastogi, V.K.; Shah, S.S.; DeFrank, J.J.; Leblanc, R.M. Infrared reflection-absorption spectroscopy and polarization-modulated infrared reflection-absorption spectroscopy studies of the organophoshorus aicd anhydrolase langmuir monolayer. *J. Phys. Chem. B* **2008**, *112*, 5250–5256. [CrossRef] [PubMed]
19. Monyoncho, E.; Zamlynny, V.; Woo, T.K.; Baranova, E.A. The utility of polarization modulation infrared reflection absorption spectroscopy (PM-IRRAS) in surface and in situ studies: New data processing and presentation approach. *Analytics* **2018**, *143*, 2563–2573. [CrossRef]
20. Baldassarre, M.; Barth, A. Pushing the detection limit of infrared spectroscopy for structural analysis of dilute protein samples. *Analytics* **2014**, *139*, 5393–5399. [CrossRef]
21. Andersen, A.; Yamada, S.; Pramodkumar, E.; Andresen, T.L.; Boisen, A.; Schmid, S. Nanomechanical IR spectroscopy for fast analysis of liquid-dispersed engineered nanomaterials. *Sensors Actuators B Chem.* **2016**, *233*, 667–673. [CrossRef]
22. Larsen, T.; Schmid, S.; Villanueva, L.G.; Boisen, A. Photothermal Analysis of Individual Nanoparticulate Samples Using Micromechanical Resonators. *ACS Nano* **2013**, *7*, 6188–6193. [CrossRef]
23. Ko, D.; Kim, S.; Jin, Z.; Shin, S.; Lee, S.Y.; Min, Y.-S. A Novel Chemical Route to Atomic Layer Deposition of ZnS Thin Film from Diethylzinc and 1,5-Pentanedithiol. *Bull. Korean Chem. Soc.* **2017**, *38*, 696–699. [CrossRef]
24. Włodarski, M.; Chodorow, U.; Jóźwiak, S.; Putkonen, M.; Durejko, T.; Sajavaara, T.; Norek, M. Structural and Optical Characterization of ZnS Ultrathin Films Prepared by Low-Temperature ALD from Diethylzinc and 1.5-Pentanedithiol after Various Annealing Treatments. *Materials* **2019**, *12*, 3212. [CrossRef]
25. Horcas, I.; Fernández, R.; Gomez-Rodriguez, J.M.; Colchero, J.W.; Gómez-Herrero, J.W.; Baro, A.M. WSXM: A software for scanning probe microscopy and a tool for nanotechnology. *Rev. Sci. Instruments* **2007**, *78*, 13705. [CrossRef] [PubMed]

26. Harding, D.R.; Goodrich, H.; Caveglia, A.; Anthamatten, M. Effect of temperature and volume on the tensile and adhesive properties of photocurable resins. *J. Polym. Sci. Part B Polym. Phys.* **2014**, *52*, 936–945. [CrossRef]
27. Ofoegbu, S.U.; Fernandes, F.A.O.; Pereira, A.B. The Sealing Step in Aluminum Anodizing: A Focus on Sustainable Strategies for Enhancing Both Energy Efficiency and Corrosion Resistance. *Coatings* **2020**, *10*, 226. [CrossRef]
28. Xu, Y.; Thompson, G.; Wood, G.; Bethune, B. Anion incorporation and migration during barrier film formation on aluminium. *Corros. Sci.* **1987**, *27*, 83–102. [CrossRef]
29. Qu, H.; Cao, L.; Su, G.; Liu, W.; Gao, R.; Xia, C.; Qin, J. Silica-coated ZnS quantum dots as fluorescent probes for the sensitive detection of Pb2+ ions. *J. Nanoparticle Res.* **2014**, *16*, 2762. [CrossRef]
30. Kharazmi, A.; Faraji, N.; Hussin, R.M.; Saion, E.; Yunus, W.M.M.; Behzad, K. Structural, optical, opto-thermal and thermal properties of ZnS–PVA nanofluids synthesized through a radiolytic approach. *Beilstein J. Nanotechnol.* **2015**, *6*, 529–536. [CrossRef]
31. Xaba, T.; Moloto, M.J.; Al-Shakban, M.; Malik, M.A.; O'Brien, P.; Moloto, M.J. The influences of the concentrations of "green capping agents" as stabilizers and of ammonia as an activator in the synthesis of ZnS nanoparticles and their polymer nanocomposites. *Green Process. Synth.* **2017**, *6*, 173–182. [CrossRef]
32. Shanmugam, N.; Cholan, S.; Viruthagiri, G.; Gobi, R.; Kannadasan, N. Synthesis and characterization of Ce3+-doped flowerlike ZnS nanorods. *Appl. Nanosci.* **2013**, *4*, 359–365. [CrossRef]
33. Estévez-Hernández, O.; Hernandez, M.P.P.; Farías, M.H.; Rodríguez-Hernández, J.; Gonzalez, M.M.; Reguera, E. Effect of Co-Doping on the Structural, Electronic and Magnetic Properties of CoxZn1−xO Nanoparticles. *Mater. Focus* **2017**, *6*, 371–381. [CrossRef]
34. ChemicalBook. Available online: https://www.chemicalbook.com/SpectrumEN_928-98-3_IR1.htm (accessed on 12 January 2020).
35. Öztürk, N.; Çırak, Ç.; Bahçeli, S. FT-IR Spectroscopic Study of 1,5-Pentanedithiol and 1,6-Hexanedithiol Adsorbed on NaA, CaA and NaY Zeolites. *Zeitschrift für Naturforschung A* **2014**, *60*, 633–636. [CrossRef]
36. Segala, K.; Dutra, R.L.; Franco, C.V.; Pereira, A.S.; Trindade, T. In Situ and Ex Situ Preparations of ZnO/Poly-{trans-[RuCl2(vpy)4]/styrene} Nanocomposites. *J. Braz. Chem. Soc.* **2010**, *21*, 1986–1991. [CrossRef]
37. Hosseini, S.A.; Mashaykhi, S.; Babaei, S. Graphene oxide/zinc oxide nanocomposite: A superior adsorbent for removal of methylene blue—Statistical analysis by response surface methodology (RSM). *South Afr. J. Chem.* **2016**, *69*, 105–112. [CrossRef]
38. Gärd, R.; Sun, Z.-X.; Forsling, W. FT-IR and FT-Raman Studies of Colloidal ZnS. *J. Colloid Interface Sci.* **1995**, *169*, 393–399. [CrossRef]
39. Thottoli, A.K.; Achuthanunni, A.K. Effect of polyvinyl alcohol concentration on the ZnS nanoparticles and wet chemical synthesis of wurtzite ZnS nanoparticles. *J. Nanostructure Chem.* **2013**, *3*, 31. [CrossRef]
40. Pandey, B.K.; Sukla, A.; Sinha, A.K.; Gopal, R. Synthesis and Characterization of Cobalt Oxalate Nanomaterial for Li-Ion Battery. *Mater. Focus* **2015**, *4*, 333–337. [CrossRef]
41. Mote, V.D.; Purushotham, Y.; Dole, B.N. Structural, morphological and optical properties of Mn doped ZnS nanocrystals. *Cerâmica* **2013**, *59*, 614–619. [CrossRef]
42. Chen, Y.; Kim, M.; Lian, G.; Johnson, M.B.; Peng, X. Side Reactions in Controlling the Quality, Yield, and Stability of High Quality Colloidal Nanocrystals. *J. Am. Chem. Soc.* **2005**, *127*, 13331–13337. [CrossRef]
43. Zenobi, M.C.; Luengo, C.; Avena, M.J.; Rueda, E.H. An ATR-FTIR study of different phosphonic acids in aqueous solution. *Spectrochim. Acta Part A Mol. Biomol. Spectrosc.* **2008**, *70*, 270–276. [CrossRef]
44. Gong, W. A real time in situ ATR-FTIR spectroscopic study of linear phosphate adsorption on titania surfaces. *Int. J. Miner. Process.* **2001**, *63*, 147–165. [CrossRef]
45. Wang, L.; Wu, Y.; Chen, F.; Yang, X. Photocatalytic enhancement of Mg-doped ZnO nanocrystals hybridized with reduced graphene oxide sheets. *Prog. Nat. Sci.* **2014**, *24*, 6–12. [CrossRef]
46. Onija, O.; Borodi, G.; Kacso, I.; Pop, M.N.; Dadarlat, D.; Bratu, I.; Jumate, N.; Lazar, M.D. Preparation and characterization of urea-oxalic acid solid form. *Process. Isotopes Mol. PIM* **2012**, *35*, 35–38. [CrossRef]
47. Pan, Y.-T.; Wang, D.-Y. Fabrication of low-fire-hazard flexible poly (vinyl chloride) via reutilization of heavy metal biosorbents. *J. Hazard. Mater.* **2017**, *339*, 143–153. [CrossRef] [PubMed]

48. Varghese, M.; Jochan, J.; Sabu, J.; Varughese, P.A.; Abraham, K.E. Spectral properties of cadmium malonate crystals grown in hydrosilica gel. *Ind. J. Pure Appl. Phys.* **2009**, *47*, 691–695.
49. de Azevedo, W.M.; de Oliveira, G.B.; da Silva, E.F., Jr.; Khoury, H.J.; Oliveira de Jesus, E.F. Highly sensitive thermoluminescent carbon doped nanoporous aluminum oxide detectors. *Radiat. Protect. Dosimet.* **2006**, *119*, 201–205. [CrossRef] [PubMed]

© 2020 by the authors. Licensee MDPI, Basel, Switzerland. This article is an open access article distributed under the terms and conditions of the Creative Commons Attribution (CC BY) license (http://creativecommons.org/licenses/by/4.0/).

Article

On-Aluminum and Barrier Anodic Oxide: Meeting the Challenges of Chemical Dissolution Rate in Various Acids and Solutions

Alexander Poznyak [1], Andrei Pligovka [2,*], Ulyana Turavets [2] and Małgorzata Norek [3]

[1] Department of Electronic Technology and Engineering, Belarusian State University of Informatics and Radioelectronics, 6 Brovki Str., 220013 Minsk, Belarus; poznyak@bsuir.by

[2] Research and Development Laboratory 4.10 "Nanotechnologies", Belarusian State University of Informatics and Radioelectronics, 6 Brovki Str., 220013 Minsk, Belarus; websulya@gmail.com

[3] Institute of Materials Science and Engineering, Faculty of Advanced Technologies and Chemistry, Military University of Technology, 2 Kaliskiego Str., 00-908 Warsaw, Poland; malgorzata.norek@wat.edu.pl

* Correspondence: pligovka@bsuir.by; Tel.: +375-44-730-95-81; Fax: +375-17-293-23-56

Received: 26 August 2020; Accepted: 8 September 2020; Published: 10 September 2020

Abstract: The chemical dissolution—in 0.1 M solutions of phosphoric, malonic, citric, sulfosalicylic, and tartaric acids and 0.6 M solutions of sulfuric, oxalic, malonic, phosphoric, tartaric, and citric acids—of aluminum (Al) and its barrier anodic oxide, with thicknesses of 240 and 350 nm, produced during the anodization of Al deposited on a sitall substrate and Al foil, respectively, in a 1% citric acid aqueous solution, was investigated. Signs of chemical dissolution for 0.1 M phosphoric acid solution and 0.6 M concentrations of all the listed solutions were found. It was shown that the dissolution rate and the nature of its change depend on the acid nature, the state of the sample surface, and the classification of the electrolytes according to their degrees of aggressiveness with respect to aluminum.

Keywords: phosphoric acid; sulfuric acid; sulfosalicylic acid; oxalic acid; malonic acid; tartaric acid; citric acid; 0.1 and 0.6 molar solution; porous anodic alumina; anodizing

1. Introduction

The properties of metal oxides and their behavior in various environments are the subject of consideration of both large review articles [1,2] and numerous publications in scientific periodicals. The research of the behavior of aluminum and its anodic oxide (AOA) in various environments is of great practical importance, since aluminum is used as a structural material and its corrosion resistance is substantially important [3,4]; aluminum has long been used for the manufacture of electrolytic (oxide) capacitors [5] and the production of microelectronic products [6,7]. The processes are studied, and methods of the electrochemical dissolution of barrier anodic oxide on aluminum (BAOA) in the composition of por-AOA (porous anodic oxide on aluminum) [8,9] are developed. Information on AOA chemical dissolution serves as the basis for carrying out assistant operations with its participation [10–12]. Por-AOA, the formation of which occurs due to the course of dissolution processes, is a matrix or template for the formation of various objects of nanotechnology [13–15]. In addition, investigations of the aluminum dissolution during anodization in combination with other results, supplementing the data on the processes of AOA nucleation and development [16–18], are of fundamental importance and have not yet been fully resolved.

Traditional models of AOA formation assumed the almost complete absence of the dissolution of the formed oxide in the case of the BAOA formation and its dissolution in the case of the por-AOA formation [19,20]. In the second case (the formation of por-AOA), at the beginning of the porous anodization process, at the stage of the linear growth of the anodic potential, dissolution does not occur at all, the BAOA is formed, and dissolution begins simultaneously with a slowdown in the growth of the anodic voltage and continues further throughout the anodization time. It was also believed that there are so-called non-dissolving electrolytes (solutions of boric acid (BA), borates, or tartrates), where BAOA grows, and dissolving ones, in which por-AOA is formed (e.g., sulfuric acid (SA) solution) [18]. However, systematic research of the initial anodization stages, undertaken by Surganov and his co-workers, showed that during porous anodization at the substantially initial moments of the anodic process—literally, in the first seconds after its beginning—the nucleation of pores occurs [21,22], accompanied by a substantially intense dissolution of aluminum [23,24], which significantly contradicted earlier models [17] and thus supplemented the understanding of the mechanisms of AOA formation.

To obtain even more convincing results, the decision was made to simulate the stage of the linear growth of the anodic potential during galvanostatic anodization using a potentiodynamic mode, setting the required voltage growth rate—"stretching", thereby, the initial stage of the anodizing process—to certain time limits to make more detailed consideration possible. As a result, specific values of the rates (current densities) of dissolution were found for 0.1 M solutions of inorganic (BA and phosphoric acid (PA)) [25,26] and organic (sulfosalicylic (SSA), oxalic (OA), malonic (MA), tartaric (TA), and citric (CA)) acids [27,28]. This concentration was chosen to increase the credibility of the investigations. Por-AOA is not formed in BA under any conditions; in the solutions of most other acids at this concentration, BAOA is formed. Nevertheless, under these conditions, a significantly noticeable dissolution of aluminum occurs. Discussing the phenomena occurring at the AOA–electrolyte interface, the authors in the publications [23–28] often used the term "dissolution accelerated by an electric field" or "dissolution accelerated by the flow of an electric current", without making attempts to consistently find out how much the dissolution rate changed during the process compared to purely chemical dissolution in the absence of anodic polarization. The relevance of this research stems from the relevance of electrochemical dissolution research. The features of electrochemical dissolution in different solutions and modes are important for understanding the mechanisms of nucleation and formation of the por-AOA cellular-porous structure. Meanwhile, the study of chemical dissolution should have shown a significant difference in the rates of these processes and dependence on the nature of the electrolyte. Previously, the authors showed [23–28] that the etching of the BAOA formed at the initial stages was due to non-trivial chemical dissolution, namely, the flow of electric current through the formed oxide film. One of the highlights of the work [23–28] and the proposed investigation that logically follows is that the aluminum dissolution was determined directly from its accumulation in the electrolyte and not assessed indirectly: gravimetrically [4], using the technique of reanodization [29,30], capacitive and impedance measurements [5,8,31], and other methods [32], which increases the reliability of the research carried out here. The second significant feature of this investigation is that the object of the research was the "real" BAOA and not the BAOA in the composition of the por-AOA, as, for example, in [8–12,29,30].

Thus, in this work, which has a fundamental novelty, as shown above, the rate of aluminum chemical dissolution and BAOA dissolution in electrolyte solutions often used in anodizing but without the application of the anodic potential was investigated.

Assumptions were made that during the experiment, the following did not occur:

a   A significant accumulation of ions Al^{3+};
b   Significant changes in acid concentrations.

In order to have grounds for such assumptions, when setting up an experiment, the solution volume for the longer dissolution of samples was increased in proportion to the exposure time T for the aluminum in solution.

2. Materials and Methods

In the experiment, the dissolution of aluminum and its alumina oxide in various molar concentrations of acid aqueous solutions was researched. The used dissolving environments were 0.1 M solutions of PA, MA, CA, SSA, and TA and 0.6 M solutions of SA, OA, MA, PA, TA, and CA, which were supplied by the Belaquilion additional-liability company and manufactured by Sigma-Aldrich, Inc. The experimental samples for the investigation of dissolution in a series of 0.1 molar acids were 99.99% pure aluminum deposited by vacuum electron beam evaporation on CT–50–1 sitall substrates. The thickness of the deposited metal was about 1 µm. The application of various kinds of coatings implies preliminary surface treatment [33]; in the mentioned case, all the samples were degreased in acetone for 15 min with ultrasonic cleaning at room temperature (about 298 K). The BAOA used in the experiment was obtained by the potentiodynamic anodizing of Al in 1% CA with an anodic voltage sweep speed of 2 V·s^{-1} to 200 V, followed by exposure at 200 V for 5 min. The thickness of the BAOA obtained was at least 240 nm. The calculation of thickness was performed based on the empirical ratio:

$$h_{BAOA} = k \times U_a \quad (1)$$

where h_{BAOA} is the thickness of the BAOA, nm; k is an empirical coefficient numerically equal to 1.2–1.4 nm·V^{-1} [34–36]; and U_a is the anodizing voltage.

The experimental samples for the dissolution investigation in a series of 0.6 M solutions were made of 99.99% polished aluminum foil of 10 micrometer thickness. Samples with a BAOA layer were obtained by anodizing aluminum foil in a 1% aqueous solution of CA with an anodic voltage scan rate of 2 V·s^{-1}, followed by anodizing in potentiostatic mode at 290 V for 10 min. The thickness of the BAOA was at least 350 nm. To investigate dissolution, samples of unoxidized aluminum and BAOA coated on aluminum were immersed in electrolyte solutions, and after a certain time, samples of these solutions were analyzed for the contents of aluminum ions present in them by atomic-emission-plasma spectrometry (ICP-AES) using a Plasma-100 spectrophotometer with inductively coupled plasma (Thermo Fisher Scientific Inc., Waltham, MA, USA), determining the concentration of Al by the emission intensity at a wavelength of 396.15 nm. The amount of dissolved aluminum was determined relative to the reacted surface area of the samples according to the following formula:

$$m_s = C_{Al} \times V_e \times S^{-1} \quad (2)$$

where m_s is specific gravity of the dissolved aluminum, µg·cm^{-2}; C_{Al} is the aluminum concentration in the electrolyte, µg·cm^{-3}; V_e is the electrolyte volume, cm^3; and S is te experimental sample area, cm^2.

It should be noted that different substrates for the 0.1 and 0.6 M solutions were carefully chosen. In the first case, a 1 µm aluminum thin film on a sitall substrate was used. The sitall has a high uniformity in thickness and planarity, which, as expected, allowed the differences in dissolution rates to be established for weak solutions and acids. With a significant increase in molar concentration, by 6 times, as expected, the dissolution rate increased significantly, which was the reason for replacing the thin film with 10-times-thicker aluminum foil.

A Keysight 5751 A programmable power supply (Keysight Technologies Inc., Santa Rosa, CA, USA), controlled by homemade software written in LabVIEW via a PC and a general-purpose interface bus cable, was used as the anodizing unit. The samples were positioned in the center of the electrochemical cell, with counter electrodes at both sides of the foil and one for the sitall substrates, parallel and in equal distance to the foil (sitall), in order to obtain a homogeneous electric field distribution. The temperature of the bulk electrolyte volume during anodizing and chemical dissolution was maintained by a thermostat at 293 K and kept as constant as possible, typically within ±1 K of the set value. The volume of the electrolyte for anodizing, which was stirred by a magnetic stirrer, was 500 mL.

3. Results and Discussion

As a result of investigations, the dissolution of aluminum and its BAOA in appreciable amounts was found to occur in many aggressive environments used in this research. However, it should be noted that a slight dissolution of aluminum and especially BAOA in most 0.1 M acid solutions made the research impossible to continue in dilute solutions for two reasons. Firstly, the concentration of the dissolved aluminum was close to the background concentration for many electrolytes (especially typical for TA and CA, not differing at high purity but slightly dissolving aluminum). Secondly, the concentration of the dissolved aluminum was at the limit of detection.

The experimental results for the 0.1 M PA solution and results for a series of experiments conducted in 0.6 M aqueous acid solutions are presented in Figure 1. The dissolution rate of the unoxidized surface of the aluminum and BAOA in a 0.1 M PA solution was researched, as it turned out, for a too short a period of time T (~10 min) in order to detect the possible nonlinearity of the dependence $m_s = f(T)$. Thus, for this case, the specific gravity of the dissolved aluminum obeys the dependence:

$$m_s = A \times T \tag{3}$$

where A is empirical constant and T is the exposure time of the sample in an acid solution, min.

At the same time, for a series of 0.6 M electrolytes, such nonlinearity is easily detected upon dissolving for a time of 10^2–10^4 min. Further analysis shows that the dependencies from Figure 1 are best described by the equation:

$$m_s = a \times (a + T)^b \tag{4}$$

where a and b are empirical values, presented in Table 1.

Table 1. Empirical constants from Equation (4) for calculating the mass of dissolved aluminum vs. that for 0.6 M solutions of phosphoric, sulfuric, sulfosalicylic, oxalic, malonic, tartaric, and citric acids.

0.6 M Acids	Dissolved Al Specific Mass m_s Dependence, µg·cm^{-2}, on the Dissolution Time T, min			
	Aluminum		Barrier Anodic Oxide on Aluminum	
	a	b	a	b
Phosphoric	6.63×10^{-2}	1.22	1.53×10^{-2}	1.36
Sulfuric	9.02×10^{-2}	1.01	1.08×10^{-4}	1.72
Sulfosalicylic	4.11×10^{-2}	0.996	1.05×10^{-4}	1.58
Oxalic	4.00×10^{-2}	0.979	1.00×10^{-4}	1.55
Malonic	1.80×10^{-2}	0.965	9.99×10^{-5}	1.35
Tartaric	6.66×10^{-3}	0.906	6.05×10^{-5}	1.25
Citric	5.59×10^{-3}	0.762	1.06×10^{-6}	1.23

The values of the empirical constants A for Equations (3) were obtained as $3.99 \cdot 10^{-2}$ for aluminum and $1.39 \cdot 10^{-2}$ for BAOA. Comparing the dependency of the graphs $m_s = f(T)$ in Figure 1a,b, one can see that the dissolution rate of the unprotected metal in PA with an increase in the acid concentration by a factor of 6 increases by approximately 1.7 times, and that for BAOA, only 1.1 times. In addition, it was found that the average dissolution rate of BAOA was found to decrease from 2.1×10^{-1} to 5.6×10^{-5} µg·cm^{-2}·min^{-1} in the following sequence of acid solutions: PA–SA–SSA–OA–MA–TA–CA. The average dissolution rate of the unprotected metal in the same sequence varied from 3.3×10^{-1} to 1.0×10^{-3} µg·cm^{-2}·min^{-1}. At the same time, if the dissolution rate of the metal and oxide in PA differed by approximately 1.6 times, the unprotected aluminum dissolved 180 times faster than the one coated with BAOA. The chemical dissolution rates of the aluminum and its dense alumina oxide increased with an increasing acid dissociation constant. This trend was not a subject to PA; the dissolution rate was higher than would follow from the indicated qualitative rule, which is possibly due to the structure of the acid anion of PO_4^{3-}. The reason may also be a possible significant specific sorption of PA and its dissociated anions to varying degrees, leading, firstly, to a local increase in the concentration of PA and its dissociation products in the surface region and, secondly, to an increase in the electric field strength of the double electric layer, which leads to a shift of Equation (10) to the right.

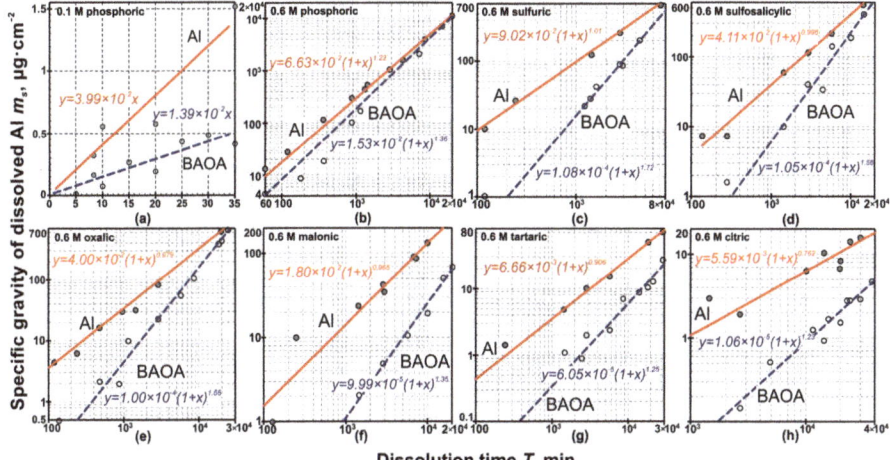

Figure 1. Dependence of the specific gravity of dissolved aluminum m_s upon exposure to a 0.1 M solution of phosphoric acid (**a**) in 0.6 M solutions of phosphoric (**b**), sulfuric (**c**), sulfosalicylic (**d**), oxalic (**e**), malonic (**f**), tartaric (**g**), and citric (**h**) acids for unoxidized aluminum (solid, red line) and aluminum protected by barrier anodic oxide on aluminum (BAOA) (dashed, blue line).

To explain the difference in the dissolution rates of aluminum and its barrier anodic oxide, as well as the nature of the dependence of the dissolution rate on the exposure time, the following model is proposed. It is assumed that in the case of both unprotected aluminum and aluminum coated with a BAOA layer, one has to take into account the oxide film formed on the Al surface (in the case of a non-anodized metal, the thickness of "natural" aluminum oxide is approximately 4.5–10.0 nm [35,37], according to the data published in [38]—20–50 Å—and in the case of a metal coated with alumina oxide, the thickness of the barrier layer was, as mentioned above, 240 and 350 nm for the experiment series in 0.1 and 0.6 M acid solutions, respectively). In this case, the specific gravity of the BAOA depends on the temperature and nature of the electrolyte from 2.69 to 3.25 g·cm^{-3} [34]. A chemical dissolution process for oxide can conditionally be described by the following equations:

$$Al_2O_3 + 3H_2O \rightarrow 2Al(OH)_3 \quad (5)$$

$$Al(OH)_3 + H^+ \rightarrow Al(OH)_2^+ + H_2O \quad (6)$$

$$Al(OH)_2^+ \rightarrow AlOOH + H^+ \quad (7)$$

The interaction with the anion, for example, Cl$^-$, can be represented as follows:

$$Al(OH)_2^+ + Cl^- \rightarrow AlOCl \cdot H_2O \quad (8)$$

However, in an acidic environment, such a continuation is possible:

$$Al(OH)_2^+ + 2H^+ \rightarrow Al^{3+} + 2H_2O \quad (9)$$

The complete reaction corresponding to the sum of Equations (5), (6) and (9) is as follows:

$$Al_2O_3 + 6H^+ \rightarrow 2Al^{3+} + 3H_2O \quad (10)$$

In the case when the solvent anion, for example, the oxalate anion, is a good complexing agent with respect to aluminum [39,40], the binding of Al^{3+} ions should be considered according to the equations

$$Al^{3+} + 2C_2O_4^{2-} \rightarrow [Al(C_2O_4)_2]^- \tag{11}$$

and, in the case of a large excess of oxalic acid,

$$Al^{3+} + 3C_2O_4^{2-} \rightarrow [Al(C_2O_4)_3]^{3-} \tag{12}$$

For the case of a PA electrolyte, the following reaction can also be considered:

$$Al^{3+} + 3H_2PO_4^- \rightarrow Al(H_2PO_4)_3 \tag{13}$$

This leads, however, to the formation of a complex compound somewhat less stable than that formed in the case of OA [39]. Reactions (5)–(13), actually removing aluminum ions from the reaction sphere, shift the equilibrium of reaction (10) to the right. On the other hand, reactions (5)–(13) lead to a decrease in the concentration of OA or PA anions, which should be especially pronounced at low acid concentrations and a small solution volume compared to the amount of metal presented.

In the interaction of Al, as well as other metals with *Red-Ox* potentials below zero, with water and aqueous solutions of acids and bases, hydrogen reduction and metal oxidation should be observed. However, for many metals (Ta, Ti, and Al), this process is kinetically inhibited to a greater or lesser extent. The reason for this potential barrier is a dense and rather inert oxide film. It is obvious that, for however long aluminum is not in the electrolyte, Al constantly remains covered by an oxide film protecting its surface. The thickness of the film, probably, in the case of an equilibrium steady-state dissolution process in each acid, is constant and depends on the nature and concentration of the acid. Thus, in the case of aluminum, both protected by a BAOA layer and coated with "natural" oxide, one should discuss dissolving a more or less thick oxide film covering the metal. Otherwise, the reaction with aqueous solutions would proceed much faster, as occurs in the case of amalgamated aluminum, which, like alkaline and alkaline-earth metals, interacts exceedingly energetically with water according to the following equation:

$$2Al + 6H_2O \rightarrow 2Al(OH)_3\downarrow + 3H_2\uparrow \tag{14}$$

Thus, the chemical dissolution process for aluminum, not just that stimulated by an electric field [41], consists of two competing reactions: the dissolution of the outer surface of the oxide film, which always covers the aluminum, and its continuous growth (restoration of thickness) to a value characteristic of each acid by the oxidation of all the new metal layers.

According to [1], the dissolution phenomena depend on the following characteristics of the oxide:

a   Covalence degrees of the oxygen–metal bond;
b   The existence of a defective lattice;
c   The anisotropy of the properties associated with the crystal orientation;
d   The presence of a film surface or bulk oxide;
e   The presence of impurities in the oxide;
f   The existence of a long-range or only short-range order.

The presence (or absence) of oxide semiconductor properties located on a metal surface also significantly affects the nature of the processes occurring during immersion. Moreover, when considering, for example, *Red-Ox* reactions occurring in a metal coated with oxide, three different cases can be distinguished, depending on the thickness of the oxide layer:

a.   The thickness of the oxide is less than 30 Å, when the oxide is sufficiently transparent for electrons due to quantum mechanical tunneling.

b. The thickness of the oxide film is more than 100 Å but less than the thickness of the space charge layer. In this case, the rate of *Red-Ox* reactions will be a function of the properties of the underlying metal, oxide, and two interfaces (metal–oxide and oxide–electrolyte).
c. The oxide layer on the metal surface has a thickness much greater than the thickness of the space charge layer. The rate of *Red-Ox* reactions will not depend on the underlying metal but only on the oxide–electrolyte interface and the semiconductive properties of the oxide.

The same trends should be expected for electrochemical dissolution. In practice, this means that, in the general case, the dissolution of a bulk (massive) oxide sample cannot be considered by analogy with the dissolution of the oxide covering the metal surface.

Furthermore, regarding oxide film thickness, its nature also affects the conductivity of the oxide. The electronic conductivity, which is important for electrochemical dissolution, is associated with the structure of the oxide and the nature of the metal–oxygen bond. The more pronounced the ionic nature of an oxide, the greater the band gap and the lower the electron conductivity. The stronger the covalency of an oxide bond, the higher the probability of *Red-Ox* reaction occurrence, and vice versa—the stronger the ionic nature of the oxide, the greater the likelihood that the dissolution includes the rate-determining ion transport across the interface of the oxide with the electrolyte.

To understand the dissolution processes, the potential distribution at the interface of the oxide with the electrolyte is important to imagine. Such a distribution, in principle, is similar to the potential distribution at the interface of the metal with the electrolyte. The electrolyte interface consists of the inner and outer Helmholtz layers and the diffuse layer.

The potential distribution near the metal–oxide interface of the sample immersed in the electrolyte solution, especially if the oxide is a semiconductor, will be more complicated than the potential distribution at the oxide–electrolyte interface. Such behavior depends on whether the substrate on which the oxide is placed is a metal, a semiconductor, or an insulator—the type and characteristics of the oxide and oxide film thickness. In the presented case, regarding the oxide layer located on the metal surface, the situation with the charge distribution will largely depend on the thickness of the film relative to the Debye length. Moreover, for a thick oxide film, when its thickness is much greater than the Debye length, the potential distribution will be similar to that for a bulk sample. If the oxide thickness is of the Debye length order, then the problem becomes more complicated and the charge distribution near the metal–oxide interface is necessary to take into account. An oxide film on a metal cannot be considered as a simple barrier to charges, both in the case of electrochemical anodic oxidation and its accompanying dissolution, and in the case of dissolution without the application of anode voltage; the properties and nature of the oxide film can be of great importance.

Moreover, one can expect the manifestation of differences in behavior between the samples (Al coated with BAOA and natural oxide layer) immersed in various electrolyte solutions, not only because of the difference in thickness but also because of the different origins of these oxide layers. A different background of "natural" and anodic aluminum oxides implies, at a minimum, the presence of various amounts of various kinds of impurities (for alumina oxide, first of all, significant amounts of electrolyte anions and/or their derivatives built into the crystal structure) and, possibly, differences in the stoichiometry and the crystalline structure of both oxides (the degree of amorphism, crystalline modification, and the presence and nature of defects).

Diggle also pointed out in his review [1] the influence of pre-treatment modes, which determine the surface under investigation and surface-structure imperfections. For example, dislocations at the crystal surface are preferred points for dissolution, since these are places of facilitated diffusion. Dislocations, due to their increased energy state, are also adsorption centers. Moreover, [1] shows the contribution of ion adsorption as well as possible complexation reactions for a prepared surface.

There is, in fact, a trivial equation for the interaction of metals with acids:

$$2Al + H_2SO_4 \rightarrow Al_2(SO_4)_3 + H_2\uparrow \tag{15}$$

This consists of several equations describing the various components of the process (which, in turn, also consist of some elementary reactions), namely:

$$Al - 3e^- \rightarrow Al^{3+} \quad (16)$$

for the oxidation of aluminum,

$$2H^+ + 2e^- \rightarrow H_2\uparrow \quad (17)$$

for the reduction of hydrogen ions present in the solution,

$$Al^{3+} + O^{2-} \rightarrow Al_2O_3 \quad (18)$$

for the formation of an oxide film on a metal, and

$$Al_2O_3 + 3H_2SO_4 \rightarrow Al_2(SO_4)_3 + 3H_2O \quad (19)$$

for the dissolution of the oxide (according to the mechanism explained by Equations (5)–(9)).

One can determine, for example, the following mechanism of aluminum oxidation, taking into account the abovementioned considerations and implying the restoration of the thickness of a continuously etched oxide film, illustrated in Figure 2. The above equations and Figure 2 show that the different stages of the total oxidation process occur near different interfaces.

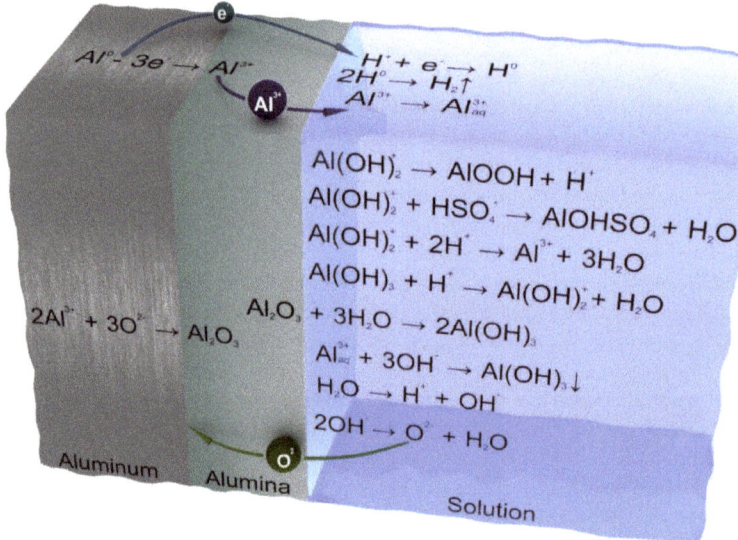

Figure 2. Schematic 3D views of aluminum's mechanism of dissolution in aqueous solutions.

It can be assumed that in acid solution, provided that its composition is constant, the oxide dissolution rate is constant and does not depend on oxide film thickness. The oxidation rate and, as a consequence, the rate of alumina formation are, ceteris paribus, a function of oxide film thickness, if only because with an increase in the aluminum oxide thickness, its resistance also increases, preventing the flow of both electronic and ion currents necessary for the process of its growth.

All the considerations mentioned above regarding the dissolution rate would be true if the considerations were dealing with an oxide sample under which an unoxidized metal would not be present. Based on the above reasoning, one can make an unambiguous assumption that the dissolution rates for non-oxidized and anodized metals should not differ. Experimental data indicate the exact opposite. For the explanation of the apparent paradox, one more possibility of a triply charged aluminum ion entering the solution should be recalled. The mentioned process is the exit of migrating Al^{3+} ions, which did not meet O^{2-} counter-ions, from the metal–oxide interface to the oxide–electrolyte interface and their subsequent exit directly into the solution, followed by hydration. It is assumed that in the case of the dissolution of aluminum without the application of anode voltage, the phenomenon of ejection is also present, i.e., the direct release of Al^{3+} ions into the solution. The assumptions about the permeability of BAOA for aluminum ions were also made in [1]. However, considering the case when a metal is located under the oxide layer, it can be assumed that the dissolution rate may depend on the thickness of the film covering the oxide precisely due to the presence of a constituent due to the direct migration of aluminum into the solution. This work proves the presence of the impermanence in time of the aluminum dissolution rate, i.e., the dependence of the dissolution rate on oxide thickness. As far as is known, the authors pointed it out in [5,42], but, firstly, these studies are of a narrowly applied character. Secondly, as a result, the dissolution of aluminum has been investigated in only one electrolyte. Thirdly, the studies are of a statement (descriptive) character, and the possible dissolution mechanisms have not been considered yet. Fourthly, an explanation of the discovered effect of the dissolution rate versus time has not been given. Since it is not clear either Al^{3+} or O^{2-} ions diffuse predominantly, Figure 2 shows both processes, and for further discussion, the specific values of the transport numbers for aluminum and oxygen are insignificant.

Then, to Equations (16)–(19), which describe the steps that occur according to the total Equation (15), one more is added that describes the migration of oxidized aluminum from the metal–oxide interface through a thin oxide layer to the oxide–solution interface with the subsequent transition to liquid phase:

$$Al^{3+}_{(metal-oxide)} \rightarrow Al^{3+}_{(oxide-solution)} \rightarrow Al^{3+}_{aq} \qquad (20)$$

Purely theoretically, one can consider the following possibility of the binding of the Al^{3+} ion, but the probability of such a process occurring in an acidic environment, and, even more so, in the presence of sufficiently strong complexing agents, is extremely low:

$$Al^{3+}_{aq} + 3OH^- \rightarrow Al(OH)_3\downarrow \qquad (21)$$

It is much more likely that the cation is bound to an insoluble complex compound with subsequent precipitation at the oxide–solution interface and, as a consequence, a further slowdown of the dissolution process.

It is substantially likely that the phenomenon described by Equation (20) provides higher dissolution rates for non-anodized metal in comparison with those for a sufficiently thick BAOA-coated metal. From this, it absolutely unambiguously follows that the counter migration of O^{2-} ions in the direction from the solution–oxide interface through the Al_2O_3 layer to the oxide–metal interface should be negligible. In the case of BAOA dissolution, the dissolution of oxide most likely occurs, at first, rather slowly but constantly accelerates as the thickness of the oxide film decreases, until the thickness of the remaining alumina oxide begins to approach its characteristic value for these conditions, while the rate of aluminum oxidation and the rate of increase (recovery) of the oxide film—due to its still-significant size and, as a result, high electrical resistance—is negligible. After the dissolution of a significant part of the BAOA thickness and when approaching the characteristic value for this solution, the metal oxidation rate begins to increase and the number of Al^{3+} ions leaving the oxide film also increases, bypassing the stage of oxide formation. It can be assumed that this characteristic thickness is of the order of the Debye length. In the case of natural oxide, its thickness is much less than the Debye length in all electrolytes, except for PA (since in PA, the voltage drop is mainly due to the double electric layer; almost nothing falls on the oxide; therefore, in PA, the natural oxide also dissolves to a thickness corresponding to the Debye length; possible reasons for the special situation of PA are mentioned above). In the case of natural oxide, the oxide grows to a thickness equal to the Debye length. Electron tunnelling becomes impossible, and the oxidation rate slows down (in PA and at the

Debye length, it is possible because the length is substantially small; aluminum rapidly oxidizes and dissolves rapidly). When, finally, alumina oxide is so etched that its thickness becomes equal to the characteristic thickness, then, whereas the dissolution rate for the oxide remains equal to that for the protected metal, the number of Al^{3+} ions entering the solution per unit time reaches a maximum and far exceeds the rate of true chemical dissolution precisely due to ion "migrants". The aluminum oxidation rate and the "migration" current, which can be defined as the charge transferred by migrating Al^{3+} ions per unit time, take on constant values. The rates of both competing processes (bleeding and growth) become maximal and equal; the process takes on a stationary and equilibrated character. In this case, the authors consider a certain idealized case, not taking into account the accumulation of Al^{3+} ions in the solvent and their possible influence on the rate of the Red-Ox reaction and the dissolution of the oxide and the change in the concentration of protons in the solution. Aluminum ions hydrated or bound into complex compounds with acid anions, by varying the dielectric constant of the solvent in the bulk and especially in the anode region, can significantly change the properties of the aggressive environment itself; the same consequences can change the concentration of H^+ ions (increase the pH of the solution).

These considerations are confirmed by the calculated values for the dissolution rate of aluminum and its BAOA obtained by differentiating the empirical expressions for specific dissolution given in Table 1. A graph illustrating the dependence of the dissolution rate for aluminum and its BAOA is shown in Figure 3. The short dashed line in the same figure shows a hypothetical one, which follows from previous arguments about the mechanisms of aluminum dissolution in acid solutions, as well as from the dependence logic $\frac{\partial m_s}{\partial T} = f(T)$, a change in the rate of the appearance of Al^{3+} ions in the electrolyte for the unstudied time domain dissolution.

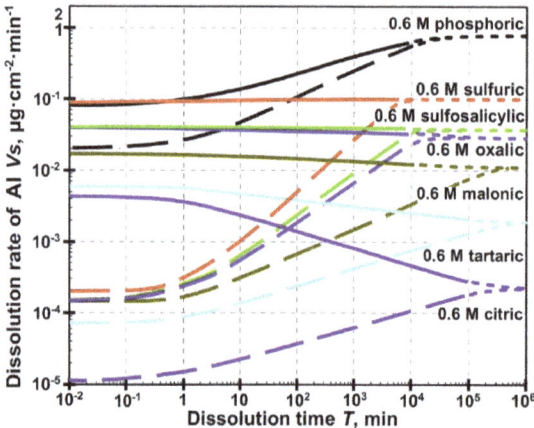

Figure 3. Dependence of the dissolution rate V_s, µg·cm^{-2}·min^{-1} for the unprotected Al (solid line) and its barrier anodic oxide layer (dashed line) on the exposure time (T, min) of the samples in acid solutions; short dashed line—extrapolation.

Based on the results of this experiment, a quantitative criterion can be proposed for assessing the solubility of acid solutions: the exponent in the expression $m_s = f(T)$ (Equation (4)) for a metal coated with atmospheric oxide. Strongly dissolving electrolytes should include a solution of PA and SA (the exponent is greater than unity, which means the characteristic, equilibrium thickness of the oxide coating the metal in the solution is less than that of atmospheric oxide). Electrolytes of medium solubility can be attributed to solutions of SSA, OA, and MA: the exponent is close to unity (0.996, 0.979, and 0.965 for solutions of SSA, OA, and MA, respectively). The thickness of the oxide coating the metal in such solutions with prolonged exposure is slightly higher than that of the "atmospheric" one.

Solutions of TA and CA should be classified as weakly soluble aluminum electrolytes. In terms of the dependence of the dissolved metal's mass on the exposure time of the sample, the exponent is already significantly less than unity; when the sample not protected by BAOA is immersed, a noticeable increase in the thickness of the oxide film occurs.

Thus, in the first approximation, it can be assumed that dissolution consists of two processes: chemical dissolution, similar to the dissolution of a massive oxide sample, the rate of which does not depend on whether the metal is under it or not but depends only on the nature and concentration of the electrolyte and oxide properties; the second process is the oxidative dissolution of aluminum, which is under the oxide layer, due to the course of migration processes and depending, in addition to the above reasons, on the thickness of the oxide. It can be reasonably assumed that the presence of metal under the oxide layer can also indirectly affect the rate of chemical dissolution. The reason for this may be the presence of a positive charge on the surface of the metal electrode, the appearance of a double electric layer, e.g., a local change in the concentration of the electrolyte in the near-electrode region.

4. Conclusions

Therefore, from the results of research on chemical dissolution—in 0.1 M solutions of phosphoric (PA), malonic (MA), citric (CA), sulfosalicylic (SSA), and tartaric (TA) acids and 0.6 M solutions of sulfuric (SA) and oxalic (OA) acids and MA, PA, TA, and CA—of aluminum and its barrier anodic oxide (BAOA), with thicknesses of 240 and 350 nm, produced during the anodization of Al deposited on a sitall substrate and Al foil, respectively, in a 1% CA aqueous solution, the following was established:

a. Upon the exposure of aluminum coated with both natural oxide and a BAOA layer with a thickness of 240 nm for 35 min in 0.1 M PA, SSA, MA, TA, and CA in noticeable amounts, Al appears only in the first solution; accumulation occurs according to a linear law.

b. With prolonged exposure ($T = 10^2$–10^4 min) of such aluminum samples coated with both natural oxide and a BAOA with a thickness of 350 nm in 0.6 M SA, PA, SSA, OA, MA, TA, and CA, an increase in the concentration of Al in solution obeys a power-law dependence.

c. The dissolution rate and the nature of its change in 0.6 M acid solutions depends on the nature of the acid and the state of the sample surface. Thus, in PA and SA, the dissolution rate for all samples increases with time; the dissolution rate for an unoxidized sample in an SSA is constant and increases with a layer of BAOA. The remaining solutions are characterized by a decrease in the dissolution rate of aluminum coated with a layer of natural oxide and an increase for samples with BAOA.

d. For all the 0.6 M electrolytes, without exception, there was a convergence over time of the dissolution rates of the unoxidized metal and aluminum protected by a BAOA layer until the oxide layer reached a thickness characteristic of these conditions. Furthermore, the dissolution rate did not change over time.

e. A classification of electrolytes according to their degree of aggressiveness with respect to aluminum is proposed. The dissolution ability criterion is the value of the degree of empirical dependence $m_s = f(T)$ for a metal that is not protected by BAOA.

Author Contributions: Conceptualization, methodology and investigation, A.P. (Alexander Poznyak); software, U.T.; validation and formal analysis, A.P. (Andrei Pligovka) and M.N.; resources, A.P. (Andrei Pligovka); data curation, A.P. (Andrei Pligovka) and A.P. (Alexander Poznyak); writing—original draft preparation, A.P. (Andrei Pligovka) and A.P. (Alexander Poznyak); writing—review and editing, A.P. (Andrei Pligovka) and M.N.; visualization, U.T. and A.P. (Andrei Pligovka); supervision, A.P. (Andrei Pligovka) and M.N.; project administration, A.P. (Alexander Poznyak) and A.P. (Andrei Pligovka); funding acquisition, M.N. and A.P. (Andrei Pligovka). All authors have read and agreed to the published version of the manuscript.

Funding: This research was funded by the state research program of the Republic of Belarus, grant "Convergence 2020" task 3.03, and the Belarusian Republican Foundation for Fundamental Research, grant No. Т20ПТИ-006.

Acknowledgments: The authors are grateful to Laryn Tsimafei and Viktoryia Karzhaneuskaya from BSUIR for help in preparing the manuscript.

Conflicts of Interest: The authors declare no conflict of interest.

References

1. Diggle, J. *Oxides and Oxide Films*; Marcel Dekker: New York, NY, USA, 1973; Volume 2, p. 281.
2. Blesa, M.A.; Morando, P.J.; Regazzoni, A.E. *Chemical Dissolution of Metal Oxides*; CRC Press: Boca Raton, FL, USA, 1994; p. 401.
3. Lemieux, E.; Hartt, W.H.; Lucas, K.E. A Critical Review of Aluminum Anode Activation, Dissolution Mechanisms and Performance. In *2001 NACE Conference Papers, Proceedings of the CORROSION 2001, Houston, TX, USA, 11–16 March 2001*; NACE International: Huston, TX, USA, 2001; p. 01509.
4. Bensalah, W.; Feki, M.; Wery, M.; Ayedia, H.F. Chemical dissolution resistance of anodic oxide layers formed on aluminum. *Trans. Nonferrous Met. Soc. China* **2001**, *21*, 1673–1679. [CrossRef]
5. Alwitt, R.S.; Hills, R.G. The Chemistry of Failure of Aluminum Electrolytic Capacitors. *IEEE Trans. Parts Mater. Packag.* **1965**, *1*, 28–34. [CrossRef]
6. Buehler, M.; Miller, A.; Andryushchenko, T. Reducing Aluminum Dissolution in High pH Solutions. U.S. Patent 2007/0152252 A1, 5 July 2007.
7. Andryushchenko, T.; Miller, A. Chemical Dissolution of Barrier and Adhesion Layers. U.S. Patent 7,560,380 B2, 14 July 2009.
8. Kim, Y.-S.; Pyun, S.-I.; Moon, S.-M.; Kim, J.-D. The effects of applied potential and pH on the electrochemical dissolution of barrier layer in porous anodic oxide film on pure aluminium. *Corros. Sci.* **1996**, *38*, 329–336. [CrossRef]
9. Santos, A.; Vojkuvka, L.; Pallarés, J.; Ferré-Borrull, J.; Marsal, L.F. In situ electrochemical dissolution of the oxide barrier layer of porous anodic alumina fabricated by hard anodization. *J. Electroanal. Chem.* **2009**, *632*, 139–142. [CrossRef]
10. Zhou, B.; Ramirez, F.W. Kinetics and Modeling of Wet Etching of Aluminum Oxide by Warm Phosphoric Acid. *J. Electrochem. Soc.* **1996**, *143*, 619–623. [CrossRef]
11. Brevnov, D.A.; Rama Rao, G.V.; López, G.P.; Atanassov, P.B. Dynamics and temperature dependence of etching processes of porous and barrier aluminum oxide layers. *Electrochim. Acta* **2004**, *49*, 2487–2494. [CrossRef]
12. Han, H.; Park, S.-J.; Jang, J.S.; Ryu, H.; Kim, K.J.; Baik, S.; Lee, W. In Situ Determination of the Pore Opening Point during Wet-Chemical Etching of the Barrier Layer of Porous Anodic Aluminum Oxide: Nonuniform Impurity Distributionin Anodic Oxide. *ACS Appl. Mater. Interfaces* **2013**, *5*, 3441–3448. [CrossRef]
13. Pligovka, A.N.; Luferov, A.N.; Nosik, R.F.; Mozalev, A.M. Dielectric characteristics of thin film capacitors based on anodized Al/Ta layers. *Proc. Int. Crimean Conf. Microw. Telecommun. Technol. (CriMiCo)* **2010**, 880–881. [CrossRef]
14. Gorokh, G.; Pligovka, A.; Lozovenko, A. Columnar Niobium Oxide Nanostructures: Mechanism of Formation, Microstructure, and Electrophysical Properties. *Tech. Phys.* **2019**, *64*, 1657–1665. [CrossRef]
15. Pligovka, A.; Lazavenka, A.; Gorokh, G. Anodic Niobia Column-like 3-D Nanostructures for Semiconductor Devices. *IEEE Trans. Nanotechnol.* **2019**, *18*, 790–797. [CrossRef]
16. Surganov, V.; Gorokh, G.; Mozalev, A.; Poznyak, A. Growth and dissolution of anodic aluminum oxide in oxalic acid solutions. *Prot. Met.* **1991**, *27*, 104–106.
17. Surganov, V.F.; Poznyak, A.A. Dissolution of aluminum in its anodizing in malonic acid solution. *Russ. J. Appl. Chem.* **2000**, *73*, 232–234.
18. Oh, J.; Thompson, C. The role of electric field in pore formation during aluminum anodization. *Electrochim. Acta* **2001**, *56*, 4044–4051. [CrossRef]
19. Young, L. *Anodic Oxide Films*; Academic Press: New York, NY, USA, 1961; p. 377. [CrossRef]
20. Henley, V.F. Anodic Oxidation of Aluminium and Its Alloys. *Pergamon* **1982**. [CrossRef]
21. Surganov, V.; Morgen, P.; Nielsen, J.G.; Gorokh, G.; Mozalev, A. Study of the initial stage of aluminium anodization in malonic acid solution. *Electrochim. Acta* **1987**, *32*, 1125–1127. [CrossRef]
22. Surganov, V.F.; Gorokh, G.G. Anodic oxide cellular structure formation on aluminum films in tartaric acid electrolyte. *Mater. Lett.* **1993**, *17*, 121–124. [CrossRef]
23. Surganov, V.F.; Gorokh, G.G.; Poznyak, A.A. Atomic-Emission Plasma Spectrometry Applied to the Initial Stages of Aluminum Anodization in Malonic Acid. *J. Appl. Chem. USSR* **1988**, *61*, 1820–1821.

24. Surganov, V.F.; Poznyak, A.A. Aluminum-Dissolution Kinetics during Electrochemical Anodizing in a Tartaric Acid Electrolyte. *J. Appl. Chem. USSR* **1992**, *65*, 2145–2147.
25. Surganov, V.F.; Poznyak, A.A. Dissolution of Aluminum during the Initial Stages of Anodization in Phosphoric Acid Electrolyte. *J. Appl. Chem. USSR* **1990**, *63*, 1901–1903.
26. Surganov, V.F.; Poznyak, A.A. Dissolution of aluminum in the first stage of anodic oxidation in solution of boric acid. *Russ. J. Appl. Chem.* **1997**, *70*, 404–406.
27. Surganov, V.F.; Poznyak, A.A. Dissolution of the Anodic Oxide at the Initial Stage of Aluminum Anodizing in Aqueous Solutions of Organic Acids. *J. Appl. Chem. USSR* **1989**, *62*, 2475–2477.
28. Surganov, V.F.; Poznyak, A.A. Dissolution of anodic aluminum oxide in the initial stage of anodic oxidation in aqueous solutions of tartaric and sulfosalicylic acids. *Russ. J. Appl. Chem.* **1998**, *71*, 253–256.
29. Vrublevsky, I.; Parkoun, V.; Sokol, V.; Schreckenbach, J. Analysis of chemical dissolution of the barrier layer of porous oxide on aluminum thin films using a re-anodizing technique. *Appl. Surf. Sci.* **2005**, *252*, 227–233. [CrossRef]
30. Vrublevsky, I.; Parkoun, V.; Sokol, V.; Schreckenbach, J.; Goedel, W.A. Dissolution behavior of anodic oxide films formed in sulfanic acid on aluminum. *Microchim. Acta* **2006**, *156*, 173–179. [CrossRef]
31. Diggle, J.W.; Downie, T.; Goulding, C. The dissolution of porous oxide films on aluminium. *Electrochim. Acta* **1970**, *15*, 1079–1093. [CrossRef]
32. Stein, N.; Rommelfangen, M.; Hody, V.; Johann, L.; Lecuire, J.M. In situ spectroscopic ellipsometric study of porous alumina film dissolution. *Electrochim. Acta* **2002**, *47*, 1811–1817. [CrossRef]
33. Rodríguez-Barrero, S.; Fernández-Larrinoa, J.; Azkona, I.; López de Lacalle, L.N.; Polvorosa, R. Enhanced Performance of Nanostructured Coatings for Drilling by Droplet Elimination. *Mater. Manuf. Process.* **2016**, *31*, 593–602. [CrossRef]
34. Field, D.J. Oxidation of Aluminum and Its Alloys. Aluminum Alloys. *Treatise Mater. Sci. Technol.* **1989**, *31*, 523–537. [CrossRef]
35. Wan, Y.; Wang, H.; Zhang, Y.; Wang, X.; Li, Y. Study on Anodic Oxidation and Sealing of Aluminum Alloy. *Int. J. Electrochem. Sci.* **2018**, *13*, 2175–2185. [CrossRef]
36. Dell'oca, C.J.; Pulfrey, D.L.; Young, L. Anodic Oxide Films. *Phys. Thin Film.* **1971**, 1–71. [CrossRef]
37. Scheer, R.; Schock, H.-W. Thin Film Technology. *Chalcogenide Photovolt.* **2011**, 235–275. [CrossRef]
38. Stojadinovic, S.; Vasilic, R.; Nedic, Z.; Kasalica, B.; Belca, I.; Zekovic, Lj. Photoluminescent properties of barrier anodic oxide films on aluminum. *Thin Solid Film.* **2011**, *519*, 3516–3521. [CrossRef]
39. Lurie, J. *Handbook of Analytical Chemistry*; Mir Publishers: Moscow, Russia, 1975; p. 488.
40. Ugay, Y. *General and Inorganic Chemistry*; Higher School: Moscow, Russia, 2000; p. 524. (In Russian)
41. Diggle, J.W.; Downie, T.C.; Goulding, C.W. Anodic oxide films on aluminum. *Chem. Rev.* **1969**, *69*, 365–405. [CrossRef]
42. Alwitt, R.S.; Hills, R.G. The Reaction of Aluminum Electrodes with a Glycol Borate Electrolyte. *J. Electrochem. Soc.* **1965**, *112*, 974–981. [CrossRef]

© 2020 by the authors. Licensee MDPI, Basel, Switzerland. This article is an open access article distributed under the terms and conditions of the Creative Commons Attribution (CC BY) license (http://creativecommons.org/licenses/by/4.0/).

MDPI
St. Alban-Anlage 66
4052 Basel
Switzerland
Tel. +41 61 683 77 34
Fax +41 61 302 89 18
www.mdpi.com

Coatings Editorial Office
E-mail: coatings@mdpi.com
www.mdpi.com/journal/coatings

www.ingramcontent.com/pod-product-compliance
Lightning Source LLC
LaVergne TN
LVHW070618100526
838202LV00012B/673